The Great Debasement

Also by
Craig R. Smith

Rediscovering Gold in the 21st Century:
The Complete Guide to the Next Gold Rush

Black Gold Stranglehold:
The Myth of Scarcity and the Politics of Oil
(co-authored with Jerome R. Corsi)

The Uses of Inflation:
Monetary Policy and Governance in the 21st Century

Crashing the Dollar:
How to Survive a Global Currency Collapse
(co-authored with Lowell Ponte)

Re-Making Money:
Ways to Restore America's Optimistic Golden Age
(co-authored with Lowell Ponte)

The Inflation Deception:
Six Ways Government Tricks Us...And Seven Ways to Stop It!
(co-authored with Lowell Ponte)

Also by
Lowell Ponte

The Cooling

Crashing the Dollar:
How to Survive a Global Currency Collapse
(co-authored with Craig R. Smith)

Re-Making Money:
Ways to Restore America's Optimistic Golden Age
(co-authored with Craig R. Smith)

The Inflation Deception:
Six Ways Government Tricks Us...And Seven Ways to Stop It!
(co-authored with Craig R. Smith)

The Great Debasement

The 100-Year Dying of the Dollar and How to Get America's Money Back

by

Craig R. Smith

and Lowell Ponte

Foreword by Pat Boone

Idea Factory Press
Phoenix, Arizona

The Great Debasement
The 100-Year Dying of the Dollar
and How to Get America's Money Back

Cover art by Dustin D. Brown
Editing by Ellen L. Ponte

Portions of this book originally appeared in two 2011 projects by Craig R. Smith and Lowell Ponte:
The Inflation Deception: Six Ways Government Tricks Us...And Seven Ways to Stop It!
and ***Re-Making Money: Ways to Restore America's Optimistic Golden Age***

Portions of this book originally appeared in
The Uses of Inflation: Monetary Policy and Governance in the 21st Century by Craig R. Smith

Library of Congress Data
ISBN Number 978-0-9711482-7-7
First Edition - November 2012

Idea Factory Press
13232 North 1st Avenue, Phoenix, AZ 85029
*Tel. (602) 918-3296 * Ideaman@myideafactory.net*

Updates, reviews and more are posted at
http://www.thegreatdebasement.com

Table of Contents

Dedication

To my wonderful wife and best friend
Melissa Smith, who makes me better
each day and raised our daughters
Holly and Katie to love the Lord
with all their hearts.
Also to my Pastor Tommy Barnett,
who taught me that doing the right thing
is always the right thing to do,
and to always hold onto the vision.

Foreword
by Pat Boone

*"Wealth obtained by fraud dwindles,
but the one who gathers by labor increases it."*

– Proverbs 13:11

The Framers of America's Constitution followed the Biblical standard for money, specifying that the U.S. Dollar be an honest measure of silver or gold.

So long as America kept faith with this standard, between the 1820s and the creation of the Federal Reserve Board in 1913, the purchasing power of our dollar over that time actually increased in value. American prosperity grew in part from keeping our money an honest medium of exchange and a store of reliable value trusted around the world.

In 1913 America began "The Great Debasement" of our currency, the turn from gold to today's Federal Reserve Notes, paper fiat money based only on politician promises. And as the Bible warned, in less than a century the buying power of our paper dollar has dwindled to only two cents of its 1913 value.

In this book my long-trusted friend and advisor Craig Smith and former *Reader's Digest* Roving Editor Lowell Ponte explain how this happened, and how our government continues to debase the dollar as a deceptive way to tax and control us. If this is not stopped soon, the dollar is doomed.

Craig and Lowell show how to free our nation, and protect our families, from soon-to-skyrocket inflation and debased values created by our ruling Inflatocracy – government of, by and for inflation. Each of us can help determine whether The Great Debasement will bring us to a new Dark Age or a new Golden Age of sound money, prosperity and higher values.

Pat

"By the common consent of the nations, gold and silver
are the only true measures of value.
They are the necessary regulators of trade.
I have myself no more doubt that these metals were prepared
by the Almighty for this very purpose than I have that iron and coal
were prepared for the purposes in which they are being used.
I favor a well-secured convertible paper currency.
No other can to any extent be a proper substitute for coin....

"The present incontrovertible currency [fiat Greenbacks] of the United States
was a necessity of war, but now that the war has ceased,
and the government ought not to be a borrower,
this currency should be brought up to a specie [gold and silver] standard;
and I see no way of doing it but by withdrawing a portion of it from circulation.

"I have no faith in a prosperity which was the effect
of a depreciated currency, nor can I see any safe path to tread
but that which leads to specie [gold and silver] payment.

"The extreme high prices which now prevail in the United States
are an unerring indication that this business of the country
is in an unhealthy condition. We are measuring values by a false standard.
We have a circulating medium altogether larger than is needed
for legitimate business. The excess is used in speculations....
The longer the inflation continues, the more difficult will it be
for us to get back to the solid ground of specie [gold and silver] payment,
to which we must return sooner or later.

"If Congress [fails soon to return to precious metal-backed money]
we shall have a brief period of hollow and seductive prosperity,
resulting in widespread bankruptcy and disaster.
There are other objections to the present inflation.
It is, I fear, corrupting the public morals.
It is converting the business of the country into gambling,
and seriously diminishing the labor of the country.
This is always the effect of excessive circulation."

– Hugh McCulloch
Secretary of the Treasury under
Presidents Abraham Lincoln
and Andrew Johnson.
Speech in October 1865

Introduction
by Craig R. Smith

"Knaves assure, and fools believe, that calling paper 'money' and making it tender is the way to be rich and happy; thus the national mind is kept in continual disturbance by the intrigues of wicked men for fraudulent purposes, for speculative designs."

– Richard Henry Lee
Declaration of Independence signer
First U.S. Senator from Virginia, 1789

After 100 years of deliberate debasement, the U.S. Dollar is dying.

America – whose power, prosperity and freedom have been secured by what once was the world's strongest, most trusted money – cannot long survive as a superpower after the dollar dies.

The dollar gave America the economic and hence military strength to defeat Nazi Germany, the Marxist Soviet Union and others bent on global conquest.

In a future without the U.S. Dollar, who will prevent a new Dark Age from descending on the world?

In this book we explore how American greatness and the dollar together produced the most innovative and prosperous economy on Earth.

Like a cat, the dollar has nine lives – and is now about to lose the last of them.

Even thereafter, we expect something called the dollar to return as a Zombie or as a ghost through the marvels of technology. The dollar, already dematerializing, will become a specter haunting our economy.

We reveal that the Great Depression that began in 1929, and the current Great Recession that emerged in 2008, were mere symptoms of the continuing "Great Debasement" of our money that began in 1913 – and that marks its 100th Anniversary in 2013.

During these 100 years, our politicians have deceptively expropriated so much of its value that a 2012 dollar has only two pennies of the purchasing power of a 1913 dollar.

This Great Debasement of America's money is by far the largest confiscation of wealth in world history – and you, your children and your grandchildren are the ones on the losing end of this century-long policy.

The U.S. Dollar was transformed from being as good as gold to being a mere piece of debased paper with no intrinsic value issued by a country whose debts now exceed its entire national income.

We explain how the Federal Reserve System, created in 1913 by Progressive politicians to give America an "elastic" money, has taken on such enormous and far-ranging powers that it has literally become "the Fourth Branch of government."

Today's dollar is just one of many currencies whose sinking worth comes from a government command, or "fiat," printed on it: "This Note is Legal Tender for All Debts, Public and Private."

The U.S. Dollar remains the world's Reserve Currency, the exchange medium on which global trade depends, yet it has become only a shadow of its once-respected former self.

We will look at Europe's Euro and China's Yuan, two other fiat currencies that may be rivals for the dollar's status as the world's Reserve Currency.

We explore how the Euro began as a Frankenstein currency and has become a Zombie. We explain how China invented ghost money, and how the U.S. Dollar is now turning into a dematerialized currency in the global economy.

The dollar has become a faith-based currency in which fewer and fewer people around the world believe.

The dollar's credibility depends on the "full faith and credit of the United States," which past generations in all nations respected.
Today's America, however, is being transformed into a European-style welfare state – a slacker nation whose spendaholic politicians keep bor-

rowing approximately 40 cents of every dollar the Federal Government now spends.

Our nation keeps falling deeper into debt by between $47,000 and $58,000 every second (depending on which ever-changing official numbers we believe), every tick of our economy's doomsday clock.

The most optimistic economists say that, at least in theory, we could pay off our current debts with steady real economic growth of as little as 4 percent per year for the next decade and more.

What such optimists leave unmentioned is that real everyday price inflation is now running at 7 percent or more per year, devouring America's paltry "stall speed" economic growth of 1.3 percent (Second Quarter 2012, adjusted as of September 27, 2012) and pulling us down into red ink in a government-unacknowledged recession of minus 5 percent annual growth.

To achieve 4 percent growth on top of such deliberately-created inflation, the politicians would need to stop increasing what government spends – don't hold your breath waiting for that to happen – *and* the economy would need to grow by at least 11 percent without recession or slowdown for the next 10 to 20 years.

The truth is that America never escaped the Great Recession that began in 2007-2008, or the Great Depression that began in 1929.

In this book we take a hard look at the cliff edge where the dollar teeters today, seeing it both from the ivory towers of economic experts and from the hard real world experience of one struggling American family.

We met this family, Ryan and Peggy Jones and their children, in our book *The Inflation Deception: Six Ways Government Tricks Us...And Seven Ways to Stop It!*

We rejoin Ryan and Peggy in this book. And we meet their nephew Patrick Jones, a worldly-wise university economist, who explains in everyday terms the often-strange views of those who control America's money supply and economic policies.

"The Greatest Debasement has not just been the devaluing of our money...," says Patrick. "What the Progressives have really debased is America itself."

We explain in detail why, and how, they have done this. And we look back at the lessons of two other historic Great Debasements, in the England of King Henry VIII and the one that undermined the Ottoman Empire.

We explore how the U.S. Dollar ascended to greatness and then, in a fateful moment in our history when many Republicans and Democrats called themselves "Progressives," our government killed the dollar that had proven successful and set a new course to replace it with an impostor currency that is a prime cause of today's many economic problems.

We examine the 1912 election that in the following year launched America's "Great Debasement" through the income tax and the Federal Reserve System and its "elastic" paper currency.

We explain why Progressives created both the Federal Reserve and the income tax in 1913, and how they have worked together synergistically to transform America into a very different country from what our Founders intended.

We consider the ominous parallels between the presidential election of 1912 and disastrous year 1913 that followed, with the election of 2012 and the fiscal cliff set to devastate America's economy and dollar in 2013.

As the American economy weakens, the dollar becomes evermore vulnerable to losing its balance and falling to its doom because of either natural or political events.

We consider several of the Tipping Points of these forces – political, natural, and international – that could trigger the dollar's demise, ending the economic and political world as we have known it.

We consider what kind of Brave New World might emerge after the dollar – and the America our once-powerful dollar made possible – have both vanished from the Earth.

We look at today's cliff-edge economy through the eyes of the Jones family to understand how the Great Debasement is changing the destiny of families like yours forever.

Fasten your seat belts, dear reader. You are about to be shown things that you never knew were manipulating your and your family's future. After reading this book, you will never again see money as you do right now.

As in most historic moments of crisis and change, the end of today's U.S. Dollar will be a disaster for those who are unprepared...and an important opportunity for some who have prepared and whose eyes are open.

Many who today take the right steps to hedge against the death of the dollar, and the vast changes this will cause, could not only survive but also thrive and prosper in the transformed world now fast approaching.

Think of this book as your survival guide for the Great Debasement and, very soon, the Post-Dollar world.

Craig R. Smith

"Money is the barometer
of a society's virtue.
When you see that trading is done,
not by consent, but by compulsion –
when you see that in order
to produce, you need to obtain permission
from men who produce nothing –
when you see that money is flowing
to those who deal, not in goods, but in favors –
when you see that men get richer by graft
and by pull than by work, and your laws
don't protect you against them,
but protect them against you –
when you see corruption being rewarded
and honesty becoming a self-sacrifice –
you may know that your society is doomed."

– Ayn Rand, *Atlas Shrugged*

Part One
The Dollar's Rise and Fall

Chapter One
A View from the Cliff

"The private sector is doing fine..."

– President Barack Obama
June 8, 2012

"Believe me, the next step is a currency crisis because there will be a rejection of the dollar... a big, big event, and then your personal liberties are going to be severely threatened."

– Rep. Ron Paul

"We thought we were doing everything right, Patrick. So why has everything gone so wrong?"

Graying Baby Boomer Peggy Jones gave her husband's nephew a wistful look and then glanced at her husband Ryan.

The three were seated around Ryan's and Peggy's dining room table, the scene of many happy family gatherings.

"We invested our savings in a home the way our parents did," offered Ryan. "But our mortgage is now more than our home is worth."

"With our savings just about gone and our income falling, it really hurts every time I go to the market and see how much prices are rising," said Peggy, an edge of emotion in her voice. Ryan gently touched her arm.

"The truth is, our jobs are shaky," said Ryan. "Even if we keep them, we may never be able to retire. Tim's graduated from State with $25,000 in student loan debt and can't find a job in his field."

"And now, dear nephew," said Peggy, "you warn us that the U.S. Dollar may be doomed. What in the world are we going to do?"

"Thank Goodness we have a University economics professor in the family," said Ryan, "who can tell us why the economy is in so much trouble and what, if anything, we can do to protect ourselves."

Seven Questions

In the candlelight, Ryan's nephew still had the bright eyes of youth despite his 35 years and strands of gray in his red hair.

"If you have the time, I can explain it," said Patrick Jones as he poured a glass of red wine.

Peggy and Ryan glanced at one another and then looked expectantly at Patrick. Patrick thought for a moment, then began.

"How much do you know about modern economics?"

"I took an economics course in college," said Ryan. "We both understand the basics of how money and the economy work."

"Do you, now?" said Patrick with a broad Irish-American grin. "Then may I ask you a few exam questions?" Ryan and Peggy nodded and awaited their "test."

"Question Number One: Why does the Federal Government tax us?" asked Patrick.

"It taxes us to raise revenue, to get money to spend," said Ryan, feeling surprised by such a simple question.

“No, that’s wrong,” smiled Patrick. “The Federal Government has at least three huge reasons for taxing us, yet directly raising revenue from us is not one of them, as you’ll understand after we unfold the mystery of how economics now are run.”

“Question Number Two: Is it better to have a dollar whose value increases, or decreases?”

“I’d feel better if our money’s value increased,” said Peggy, offering a wry smile.

“No, believe it or not, according to those who control our money it’s far better to have a dollar with shrinking value,” said Patrick. “They believe it would be a disaster if our dollars grew in value, if we had deflation instead of inflation, as I’ll soon explain.”

In Vino Veritas

Seeing Peggy’s and Ryan’s puzzlement, Patrick raised his wine glass in a toast.

“*In vino veritas*,” he said. “As the ancient Romans used to say, ‘In wine there is truth,’ or at least lowered inhibitions and loosened tongues.”

“At this private family dinner, just between us,” said Patrick, “I can tell you the whole, Politically-Incorrect truth about how our economic system and money now really work....things we dare not say at the university. I can explain the strange ways that the people who control our money think about it, and why the U.S. Dollar as we have known it is near its end.”

“The truth,” said the nephew Ryan and Peggy nicknamed Professor Pat, “is that we now live in an Alice-in-Wonderland, topsy-turvy economy very different from what the textbooks and practical life experience used to teach.” Patrick set his glass on the table.

“Here’s Question Number Three, Uncle Ryan: Is risk good or bad?”

“We should try to eliminate risks,” Ryan replied.

“No, risk is good. One of the biggest reasons for today’s economic problems is that we didn't, and still don't, have enough risk," said Patrick with an impish grin. "In America we are suffering because of a risk deficit."

Cross to a Vampire

"Let me ask you a question," said Ryan, pointing to his wedding ring. "What do economists think of gold? You hear a lot about it in the news."

"Economists who plan our economy hate gold, but even more they fear it," said Patrick. "Mentioning gold to them is like holding up a crucifix to a vampire."

"You see, if we went back to gold standard money – the system America's Framers put in the Constitution – our politicians couldn't play most of the manipulative games with our money that they do today," Patrick continued. "Gold is the only genuine individually-sovereign antidote for what they've put into the body politic and body economic."

"Yet, ironically, the world's money-gaming central banks have recently rushed to acquire lots more gold for their own reserves, because they know gold's power – and know what is about to happen. This should be one of those things that make you go 'Hummmm'."

"Counting yours as Question Four, Uncle Ryan, here's Question Number Five: Is it good or bad to save your money?" asked Patrick.

"Ben Franklin said that a penny saved is a penny earned," said Ryan.

"No," said Professor Pat. "Saving is bad for the economy, at least according to those who now run it." He took another thoughtful sip from his glass as Ryan and Peggy shook their heads in bewilderment.

"Financial Repression"

"Anyway, in today's economy, saving actually costs you money," said Patrick. "In today's ruling Inflatocracy – our government of, by and for inflation – a penny saved is just a penny lost to inflation."

"The government in effect skims the real profit from your savings through a cunning deliberate process it uses that turns today's bank savers into suckers. We economists call this – again, believe it or not – 'financial repression.'"

"And here's Question Number Six: Does it matter if the Federal Government goes deep into debt?"

"Of course it does," said Peggy.

"No, it doesn't, according to some influential economists who now say that stratospheric Federal Government debt scarcely matters at all. They say that the government should just print and spend as many trillions, or tens or hundreds of trillions, of dollars as it takes to improve the economy," Patrick told his dumbstruck aunt and uncle.

"You've probably noticed that our politicians certainly keep spending as if debt doesn't matter."

The Market Paradox

"Here's Question Number Seven, our last for right now," Patrick continued. "Is it good if the economy improves?"

"I'm beginning to think that you'll tell us an improving economy is bad," said Peggy.

"Bingo," laughed Patrick. "Paradoxically, we now have a stock market that goes down on good economic news and up on bad economic news because the stock market really doesn't reflect the capitalist free market anymore, as I'll explain later. It reflects and responds to something else entirely."

"So is *everything* about the economy upside-down?" asked Ryan.

"Not everything," said Patrick, "but enough has changed that the economy and its new real rules are no longer entirely what you learned back in school. Welcome to the Brave New Economy."

The Rabbit Hole

"As I promised, I'm happy to explain all this to you," said Patrick. "I must, however, give this warning: once you learn this information, you'll never again be able to see the economy or politics as you do right now."

"As they said in the movie 'The Matrix,' if you decide to take the blue pill you'll wake up in your bed tomorrow and everything will seem just as confusing and nonsensical as before."

"If you take the red pill, your eyes will open. As the character Morpheus says in 'The Matrix,' you'll stay in Wonderland, and I will show you how deep the rabbit hole goes."

"Do you really want to understand why so much has gone wrong with America's economy and done so much to hit you and Aunt Peggy?" Ryan nodded.

"Then let's start with the only two things we all know for certain – where we are, and what time it is. We know that we are always here, and we know that the time is always now. So let's begin with your here and now."

"As you've already sensed, as Morpheus says, 'like a splinter in your mind,' you and I and the country feel foreboding, that we are standing at the brink of a cliff, and it's a long way down," said Patrick. "I hope you're not scared of heights."

Patrick raised his wine glass in another toast.

"Here's to the U.S. Dollar – what's left of it. Alas, it's now too small to be either our hang-glider or our parachute. Alas, it used to be praise to say something was as 'Sound as a Dollar,' yet now it's almost an insult."

"Run on the Dollar"

"So, Professor Pat, what caused the Great Recession that hit us in 2008?" asked Ryan.

"It had an immediate trigger, an attack on U.S. financial institutions from overseas, as I'll later explain," Patrick replied. "It had a broader cause in the collapse of the world market in bad mortgage bundles and the popping of the housing bubble."

"Yet if you look at almost any of today's economic problems, they have been caused ultimately by the Great Debasement of our money that the Progressives began in 1913," said Patrick. "Tonight I shall, as we academics say, 'unpack' exactly how this Great Debasement has been done and the destruction it keeps causing."

"To answer your question, Uncle, the Great Debasement has greatly weakened the U.S. Dollar. The weak dollar caused the Great Recession."

"During the years around 2005-2006, housing and fuel and commodity prices were soaring," said Patrick. "Much of the reason for this, as Harvard Ph.D. Brian Domitrovic theorized in a 2012 *Forbes* Magazine column, was what he called a potential 'run on the dollar on account of fears of its devaluation, in spades, from 2003 to 2008' after the Federal Reserve pushed interest rates below one percent."

"This, argues Domitrovic, a professor at Sam Houston State University in Houston, Texas, caused 'major investment shifting into hard assets corresponding to fear for the dollar's soundness....The government in its role of guarantor of the currency caused this crisis....mismanaged the currency.'"

"This, as you'll soon learn, is one more example of what Nobel laureate economist Friedrich Hayek called 'that vicious circle wherein one kind of government action makes more and more government control necessary.'"

"Government causes economic sickness by interfering with the free market, and then uses this as a pretext to impose itself through authoritarian laws such as Dodd-Frank by pretending that more government is the cure," said Patrick. "It's literally like poisoning someone to death a few toxic drops at a time."

"This is much of the reason the Cato Institute study *Economic Freedom in the World 2012* shows the United States dropping from the third freest economy on Earth only a few years ago all the way down to the 18th freest in 2012."

"The Cato think tank now ranks U.S. economic freedom not only below that found in the free market city-states of Hong Kong and Singapore, but also below the high-tax European welfare states of the United Kingdom, Finland and Denmark – and even below the Persian Gulf states Qatar, the United Arab Emirates and Bahrain," said Patrick.

"Economic liberty in the United States has been plummeting under the boot of Progressive politicians who are openly hostile to capitalism. How much prosperity is possible in such a political environment?"

Ryan was dubious. "A Progressive friend tells me that the economy will work better when coordinated by central planning."

"It certainly worked well in the late Soviet Union," said Patrick sardonically. "No, Adam Smith's 'invisible hand' produces a vastly freer, more productive economy than the Progressives' visible boot of ideological, crony control. Here's what Hayek said," continued Patrick, holding up a piece of paper with these words:

"The effect of the people's agreeing that
there must be central planning,
without agreeing on the ends,
will be rather as if a group of people
were to commit themselves to take a journey together
without agreeing where they want to go;
with the result that they may all
have to make a journey which
most of them do not want at all."

– Friedrich Hayek

"That, my dear Aunt and Uncle, is the history of Progressivism in a nutshell. The Great Debasement has been taking us forcibly on a collectivist journey that few Americans, had they known, would have agreed to take. We would never have given Progressives a blank check of corrupting power."

The Cave-in

Like most Baby Boomers, Ryan and Peggy shared the American Dream of their Greatest Generation parents – a suburban home, white picket fence, roses, a yard for the kids to play in.

They trusted that their home was an excellent investment. The Lord, they were told, was not making any more land. With a fast-growing post-World War II population in America, the price of good land seemed to have no place to go but up.

The tax-deductible advantages of mortgage interest made home ownership an investment tax shelter available to middle class Americans.

This American Dream produced solid, low-crime neighborhoods of hard-working property taxpayers whose homes were their castles and their pride. It was proof that, like their pioneer ancestors, they could become landowners in America.

The Dream Turns Nightmare

What turned this dream into today's nightmare?

We explored the causes of the American housing market collapse that is still dragging down our economy in two full chapters of our book *Crashing the Dollar: How to Survive a Global Currency Collapse.*

President Jimmy Carter in 1977 wanted to make home ownership available to more Americans too poor or credit-impaired to qualify for a home mortgage.

Mr. Carter signed into law the Community Reinvestment Act (CRA), which gave community organizers and politicians the power to strong-arm banks into giving mortgages to previously unqualified high-risk borrowers.

The government denied permission for banks that failed to fulfill CRA loan quotas to do corporate mergers, acquisitions or otherwise expand their businesses.

Ninjas

The banks described many of the borrowers the CRA forced on them as "NINJAs," people with No Income, No Job or Assets, and therefore little traditional credit-worthiness.

Under great political pressure from President Carter and later President

Bill Clinton and George W. Bush, banks were coerced into making a trillion dollars of sub-prime home loans, money unlikely ever to be repaid.

This flooded the housing market with easy money, which sent home prices soaring. Television ads offered homeowners equity lines for up to 125 percent of their home's value, a gamble many lenders were eager to make when home prices were going up 20 percent every year.

By the late 1990s, inflation-adjusted average American wages had scarcely risen in more than 20 years. Yet suddenly Americans felt, and spent, like millionaires. Many turned their suddenly-expensive homes into piggy banks by borrowing their equity.

And, again, banks were pressured not to deny such loans to lower middle class applicants with weak credit. A young lawyer named Barack Obama, who would be employed by the radical activist group ACORN, helped file one of the lawsuits that further loosened lending standards.

Spreading the Risk

Many banks began selling off their sub-prime mortgages before the loans went bad as a way of shedding their own risk. Many of these mortgages were packaged into bundles that were given high credit ratings, were insured by giant corporations worldwide including AIG, and were sold to trusting investors in Iceland, Norway and around the world. These were ticking financial time bombs that soon blew economies apart.

When the housing investment bubble burst and home prices began to plummet, more than 11 million homeowners found themselves, like Ryan and Peggy, owing more on their home than it was worth. Being "underwater" took on a whole new meaning.

Many borrowers, thanks to CRA political pressure, had been provided mortgages with little or no down payment. Many who had invested nothing in their upside-down homes simply stopped paying and walked away. This further depressed the home values of those who honored their bank agreements.

Home prices remain depressed, on average, by more than 30 percent from their peak.

Approximately 25 percent of American homes are underwater, like Ryan's and Peggy's. One American home or apartment in every 11 is now vacant.

Shadow Inventory

The banks, as we explored in *Crashing the Dollar*, are reportedly holding back a "shadow inventory" of at least 1.7 million foreclosed homes they are not yet putting on the market, lest this oversupply drive down home prices even more.

The quasi-governmental lenders Fannie Mae and Freddie Mac were nationalized and have been used since 2008 to absorb the flood of toxic mortgages poisoning bank balance sheets. They reportedly now hold at least $5.3 Trillion in mortgages, with at least 13.5 percent of these sub-prime. Ultimately taxpayers could get stuck with $1-3 Trillion in real losses.

Some of the major banks involved in this mess were, for the sake of the larger economy, forced by government to accept loans, merge with other banks, and accept additional regulation and scrutiny over their lending.

"Too Big to Fail"

The giant banks deemed "too big to fail" are now 30 percent bigger than before and, thanks to a regulatory squeeze on smaller banks, have a much larger share of the market than before the 2008 economic crisis.

Government's Politically-Correct social engineering through CRA caused irresponsible lending and a market collapse. Now, as the politicians point fingers of blame at others, new heavy regulation known as Dodd-Frank has forced banks to curtail much of their traditional responsible lending.

Nearly two-thirds of Dodd-Frank rules are yet to be put in final form. The more than 8,000 pages of rules written thus far use language disturbingly close to that in the emerging new global banking rules known as Basel III. This leads some experts to suspect that we soon may see a coalescing of global debts through the International Monetary Fund (IMF), World Bank or some other global monetary authority.

These arrogant politicians and activists once boasted that they had driven home ownership to a record high 68 percent of American families. The economic nightmare they created has now driven home ownership down to its lowest level in nearly three generations.

"As of mid-2012, the Obama Administration had resumed threatening lenders with fines and other government penalties if they did not increase their lending to minority individuals who were poor credit risks," said Patrick. "So despite the nightmare it has caused, the Progressive wealth redistribution scheme of CRA continues."

In today's America, almost 30 percent of young people up to age 34 now live with their parents – the highest such percentage since just after World War II. As Congressman Paul Ryan of Wisconsin famously said, they lie in their childhood bedrooms looking at fading posters of President Obama's 2008 campaign on their walls or ceilings and remember his faded promises of hope and change.

Some have said that this is the same kind of "infantilizing" of our culture as the provision in President Obama's health plan that keeps people until age 26 on their parents' health insurance coverage. And while they postpone adulthood and having children of their own, America's fertility rate reportedly has dropped to a non-replacement level of only 1.87 births per woman, even though the official rate remains around 2.01.

Young members of Generation Y in polls now exhibit a very low interest in ever owning their own home, having seen the loss their parents suffered in the home price cave-in. This old American Dream may soon be replaced by people who rent, not own, and have more transient, less-rooted lives and values.

Marketplace Manipulation

This high-handed government interference in the marketplace, combined with massive government borrowing, has crowded out small businesses and homeowners from getting needed loans to expand, hire, build or buy.

The still-depressed housing market has crippled the economy, because at least one in six jobs is home-related – manufacturing and selling major appliances, lumber, roofing, garden supplies and thousands of other products and services for homeowners.

Over our entire history, the United States has never had sustained recovery from any recession or depression without a solid housing market recovery. With yet another wave of home foreclosures expected, the housing market and hence the whole economy could take anywhere from two to 10 years to recover.

Housing has become such a slow-healing Achilles Heel to America's hobbling economy that in September 2012 the Federal Reserve stepped in with an open-ended promise of $40 Billion every month in Quantitative Easing stimulus money targeted to buy up troubled mortgage obligations.

Depressed housing has also depressed employment. People who cannot afford to sell their homes or sacrifice their credit are unable to move to find available jobs.

Unemployment officially remained above 8 percent for the first 44 months of the Obama presidency. Real unemployment, measured to include the underemployed and discouraged workers as was done during the Great Depression before politicians started gimmicking the counting methodology, is nearly 15 percent.

"All net jobs created during the Obama administration have been part-time jobs," wrote billionaire investor and publisher of *U.S. News & World Report* Mortimer Zuckerman in July 2012, "which generally means that these workers receive no benefits and that their pay is inadequate to enter the middle class."

Fewer Americans are working today than had jobs in year 2000, noted Zuckerman, "despite the fact that our population has grown by 31 million and our labor force by 11.4 million since then."

"This is, in effect, the modern-day Depression," wrote Zuckerman in September 2012. "Roughly 15 percent of the population, a record representing over 46 million Americans, are in the food stamp program," roughly double the percentage of Americans from 1970 to 2000, with "about 400,000 people...signing up each month over the past four years."

Unemployment has been projected to get a lot worse if the politicians shove America off the economic cliff into Taxmageddon on January 1, 2013. The accounting and analysis firm Ernst & Young, for example, in July 2012 released a study projecting that if President Obama's tax increases take effect, the U.S. economy would over time lose 710,000 jobs because of these taxes. Economic output would fall by $200 Billion a year, and "real after-tax wages" would fall by 1.8 percent.

"I think Ernst & Young's estimates of trouble from these tax increases are too small," said Professor Pat, "because the tax increases if we go off the fiscal cliff are much bigger and hit more people than most folks know."

"On January 1, 2013, the top tax rates of virtually every major federal tax will increase sharply," wrote *Wall Street Journal* economist Stephen Moore in October 2012. "That is when the tax increases of Obamacare go into effect, and the [President George W.] Bush tax cuts expire, which President Obama refuses to renew for the nation's small businesses, job creators, and investors. That is the English translation of singles making over $200,000 a year, and couples making over $250,000."

If America goes off the fiscal cliff at midnight on December 31, 2012, wrote Moore, "the top two income tax rates will jump nearly 20 percent; the capital gains tax rate will soar by nearly 60 percent; the tax rate on dividends will nearly triple; the Medicare payroll tax will rocket up by nearly 62 percent for these disfavored taxpayers; and the death tax will rise from the grave with a 57 percent rate increase."

America remains nominally a capitalist country, yet under President Obama corporations pay the highest income tax rate among all major nations – an average combined federal and state corporate tax rate, Moore calculates, of nearly 40 percent.

"Even communist China has a low 25 percent corporate rate by comparison," wrote Moore. "American businesses are uncompetitive in the global economy with this tax disadvantage."

The cost of taxes is more than the amount collected, as Duke University research scholar Christopher Conover explored in a study published in 2010 by the Cato Institute. He found that the "marginal excess burden of federal taxes most likely ranges from 14 to 52 cents per dollar of tax revenue, averaging about 44 cents for all federal taxes." Conover believes that Congress should account for this excess burden of taxation when planning tax policies and passing tax laws.

In September 2012, a poll of small business owners conducted for the National Association of Manufacturers found that 67 percent said the market had too much uncertainty for them to expand, grow or hire new workers. Fully 55 percent said that they would not have started their businesses in today's economy.

Rigging the Numbers

"Even the official unemployment numbers are deceptively low for a variety of reasons, some of them political," explained Professor Pat.

The Obama Administration expanded government benefits for going back to school, which took many off the unemployment rolls.

Mr. Obama made it easier to qualify for Social Security Disability payments by claiming hard-to-prove mental problems, and millions since have ended their 99 weeks of Unemployment payments by going onto permanent Disability payments. More than eight million Americans are now on Social Security Disability, more than the entire population of New York City. This has put a huge additional drain on Social Security funds for future retirees.

Mr. Obama altered President Bill Clinton's 1996 welfare reform law that required able-bodied welfare recipients to seek work. This bipartisan reform reduced welfare rolls in America by 50 percent by making it a bit harder for slackers to turn the social safety net into a hammock. The Obama Administration circumvented this reform by allowing states to seek waivers from this requirement.

When welfare recipients stop applying for work, within four weeks they are dropped from the Bureau of Labor Statistics count of unemployed people. It meant that the final jobs report prior to the 2012 election would show fewer unemployed people, even though the Obama welfare change did not lead to the employment of a single person.

We first published this analysis in July 2012, along with our prediction that the jobless rate would suddenly fall into the 7 percent range just prior to the November election, and were interviewed about it on many media outlets.

Three months later, the government's September jobs report, issued October 5, surprised experts by saying that the jobless rate had somehow declined from 8.2 to 7.8 percent.

This was the largest one-month jobless decline in 29 years, which back then happened when the economy under President Ronald Reagan was growing by more than 5 percent. In September 2012 the most recent adjusted growth rate for the economy was an anemic 1.3 percent, with many economic indicators pointing to even further slowing. Obama Labor Secretary Hilda Solis, a former member of the Congressional Progressive Caucus, adamantly denied that the number had been manipulated.

80 is the New 65

Even for those with jobs, the average American income since January 2009 has declined by $4,019 per year, according to U.S. Census Bureau data.

For millions like Ryan and Peggy, the housing crisis has reduced their net worth – in effect, their life savings – by approximately 39 percent.

Those aged 35-44 have been hit hardest of all, losing on average 55 percent of their net worth, most of which disappeared in the declining value of their homes.

This is especially hard because people 44 and younger may receive only the remaining crumbs of a Social Security cake eaten up by older generations. Yet there will be only two of them in the workforce to pay for every one person receiving Social Security benefits. You could say that we reached this point in June 2012, when the number on Social Security or its disability benefits reached 50.54 percent of the 111,145,000 workers in the private workforce.

And, to add insult to injury, lots of Baby Boomers and Generation-Xers will not retire because they need the income, which means that fewer jobs will open for younger workers on the promotion ladder, which will make their real career average income lower than expected.

The politicians also looted $2.66 Trillion from the Social Security Trust Fund and used it to assure their re-elections over the past 20 years, leaving only IOUs that will require yet more future high taxes to replace the politically-expropriated Social Security money that taxpayers have already paid.

And to pay for his new national health plan, President Obama diverted nearly three-quarters of a trillion dollars from Medicare's funds.

Comedians used to joke that "80 is the new 65," that people soon will have to work far into old age before they can afford to retire. In 2012 the chief executive of insurance giant AIG said that, in fact, retirement age for government benefits will have to be raised to between 70 and 80. This is no longer a laughing matter.

"$58,000 Every Second"

"Above all, the U.S. Dollar is doomed because of all the debt the Great Debasement has piled onto our currency and economy," said Patrick. "That debt since President Obama's 2009 inauguration has, by one estimate, expanded by 50 percent and as of September 2012 was equivalent to $136,260 per household in America."

"During the past several years approximately 41 cents of every dollar the Federal Government has been spending is borrowed money. Our spendaholic politicians have been borrowing up to $58,000 every second."

"What this means is that every two hours we borrow as much, in inflation-adjusted dollars, as the United States once spent to buy the Louisiana Purchase plus Alaska – roughly two-thirds of all American land west of the Mississippi River."

"Let me repeat that," said Patrick, speaking slowly and emphatically. "Every two hours our government now borrows more than we once spent to BUY two-thirds of all the land west of the Mississippi. This rate of borrowing, as every non-Progressive economist agrees, is simply unsustainable for much longer."

"How big is our real debt? Think of it this way – and I won't use the more realistic high numbers, which assume the politicians will of course keep borrowing and spending. I'll use very, very conservative numbers," said Patrick.

"The immediate government debt in 2012 topped $16 Trillion and went above America's Gross Domestic Product, everything we produce in a year."

"Nearly half of this debt must be refinanced in the next three to four years," said Patrick, "and our creditors holding this debt such as the People's Republic of China or Japan – each of whom holds more than $1.1 Trillion of our Treasury obligations – could demand either payment or a higher interest rate to roll the debt over for another short term."

"Part of President Bill Clinton's economic 'success' came from refinancing U.S. debt into shorter-term obligations, which entail less long-term inflation risk for the lender and therefore carry lower interest rates," Patrick explained.

"This produced savings in the beginning, but now we have very heavy debts coming due. They are so heavy that if interest rates go up by one percent, this over a decade would cost American taxpayers an extra trillion dollars. And chances are that the rate foreign lenders will demand could be several percent higher than we pay today," said Professor Pat.

"If that happens, just paying the interest on our debt will start to crowd out every other item in our national budget, including defense spending. Soon all the politicians will pay are entitlement payments and interest."

"So is that the whole debt?" asked Peggy.

"Sadly, no, Aunt Peggy," said Patrick. "On top of this immediate $16 Trillion, which is already on the way to exceeding $20 Trillion by 2016 or sooner, is our total national debt, public and private."

"This combined debt is probably around $67 Trillion. By comparison, the entire Gross Domestic Product of the world and its 7 billion people is only around $62 Trillion, so our Progressive politicians are spending more than the whole world earns." said Patrick.

"On top of that," he continued, "is our government's 'unfunded liability' for Social Security, Medicare, government pensions and other promises made by politicians. This at the moment adds another $120 Trillion in long-term taxpayer obligations."

"And then there is the dark matter of the financial universe," said Patrick. "These are derivatives, mostly privately-made obligations by insurance companies, corporations and others whose value is derived from something else."

"We can't ignore these because up to $1.28 Quadrillion – that's a one followed by 15 zeroes – worth of them at face value are considered to be assets by the world's biggest banks and corporations. How much are they really worth if they had to be liquidated? Nobody knows."

"Oh, and I almost forgot," said Patrick, "we can't forget derivatives because some of our biggest American banks recently shifted at least $154 Trillion of derivatives from their uninsured banking arms to those insured by the Federal Deposit Insurance Corporation – thereby potentially putting taxpayers on the hook for them."

"My best conservative estimate is that American banks and corporations are holding at least $333 Trillion in derivatives," said Patrick.

"Add these together and we get a very conservative estimated total debt that someday must be paid off of approximately $536 Trillion," said Professor Pat.

Peggy covered her face with her hands.

Astronomical Debt

"Here's where it gets interesting," Patrick continued.

"Imagine taking $1 bills out of your wallet and making a stack of them. A stack one million miles tall would require roughly 14.3 Trillion $1 bills."

"A stack of our combined debt – $536 Trillion in $1 bills – would be almost 37.5 times taller – almost 37.5 million miles high," said Patrick.

"This means that paying all of America's debts would require a stack of $1 bills that could reach from Earth to the planet Mars, which at its closest approach is 35.8 million miles away."

"In fact our stack of debt money would go past Mars by 1.7 million miles, enough dollars of debt left over to build almost seven stacks of dollars between Earth and the Moon. And don't forget, this debt keeps increasing by up to another $58,000 every second."

"I think I'm going to faint," said Peggy.

"I already have," said Ryan.

"Now here's my question," smiled Patrick. "In a country where half the people pay no income tax, and where 49.1 percent of households have someone there receiving some kind of government check or benefit, does anyone seriously think that this debt will ever be paid?"

Both Ryan and Peggy shook their heads. "What can we do?" asked Peggy.

"We have two ways to end this debt," said Patrick. "One is for America to declare bankruptcy, which in countries is called defaulting on the debt."

"National governments seldom do this as long as they have taxpayers they can burden with the debt's cost – because declaring bankruptcy might not be a good option for a government devoted to borrowing. Defaulting might make it harder to find future lenders who charge less than loan-shark interest rates."

"Understandable," said Ryan.

"There's one way to 'pay' our obligations," said Patrick, "and that is by what we economists call 'monetizing the debt,' just turning on the printing presses and manufacturing out of thin air the equivalent of $536 Trillion."

The U.S. Dollar is Legal Tender, meaning that people and even the government can be forced to accept these pieces of paper as payment for any debts denominated in dollars," said Professor Pat. "I'll explain how strange this can be when we discuss what economists now call 'Modern Money.'"

"Meanwhile, back in the real world, as we all know from the law of supply and demand, in an economy flooded with $536 Trillion in paper fiat currency, the exchange value of a dollar would fall to zero, the way the German Mark did in 1923 Weimar, Germany. The dollar would simply die as its value does."

Both Ryan and Peggy shook their heads in disbelief.

"To understand where the U.S. Dollar is going, we need to know how it was born, why it rose to greatness, and how the Great Debasement has driven it down to only two pennies of its once-awesome purchasing power," said Patrick.

"Standing here on the fiscal and monetary cliff, with the dollar's and our economy's survival at stake, we can learn a lot about America – and our values – by taking a moment to discover our dollar's surprising past."

"I sincerely believe, with you, that
banking establishments are more dangerous
than standing armies,
and that the principle of
spending money to be paid by posterity,
under the name of funding,
is but swindling futurity
on a large scale."

– Thomas Jefferson
Letter to John Taylor
May 28, 1816

Chapter Two
Ascent of the Dollar

"With the exception only of the period of the gold standard, practically all governments of history have used their exclusive power to issue money to defraud and plunder the people."

– Friedrich A. Hayek
Nobel Laureate Economist

The American Eagle was born in a nest of paper dollars. America's Founders came away from this determined that a free and strong American Republic must be built on solid, not paper, money.

During the American Revolution, the cash-starved Continental Congress did what most modern governments do. Lacking sufficient money, it recruited the printing press as a weapon to conjure paper money out of thin air.

When our War for Independence began, America's 13 former colonies and their 2.4 million people had in circulation money equivalent in value to around $12 Million Spanish Dollars. New York delegate Gouverneur Morris proposed that the Congress fund the war by printing a paper Continental currency denominated in dollars.

In 1775 Congress printed $6 Million of the new currency, increasing money in the colonies by 50 percent in one year.

In 1776 the Continental Congress issued $19 Million more, then another $13 Million in 1777, $64 Million in 1778, and an astonishing $125 Million in 1779.

"There was no pledge to redeem the paper, even in the future," wrote economist Murray Rothbard in his *History of Money and Banking in the United States*, "but it was supposed to be retired in seven years by taxes levied pro rata by the separate states."

"The retirement pledge was soon forgotten," wrote Rothbard, "as Congress, enchanted by this new, seemingly costless form of revenue, escalated its emissions of fiat paper."

Wallpapering America

Within five years, these lawmakers wallpapered America with $225 Million of this unbacked paper, putting nearly 19 dollars in circulation for every one at the start of the Revolution – an increase in paper money of almost 2,000 percent.

As historian Edmund Cody Burnett observed, "such was the beginning of the 'federal trough,' one of America's most imperishable institutions."

The cheaply-printed bills were easy to counterfeit by the dishonest – and by the British, who flooded the rebel states with more millions of fake Continentals to undermine popular support for and trust in the Revolution.

The supply of these notes quickly outpaced demand, and as Economics 101 teaches, the value of these scraps of paper plummeted.

Revolutionary soldiers were paid in Continentals. When farmers or merchants refused to accept the inflated paper currency for food or supplies the Continental Army needed, their goods were sometimes "impressed" and they were forced to accept the devalued paper as payment.

When the first Continental dollars were issued, Americans were willing to exchange them for silver dollar coins at 1 or 1.5 to one. A year later it took three Continentals to buy such a coin. By December 1779 the exchange rate had fallen to 42 to one. By December 1781 a dollar coin fetched 168 of the paper bills.

Imagine the price of potatoes at your local market rising in only six years from $1 per pound to $168 per pound, and you get some idea of how soldiers and merchants felt about the value of being paid with America's first fiat paper currency.

By 1779, the Continental Congress sold the first of what eventually would be $600 Million worth of a parallel currency, loan certificates, that after the war became part of the young nation's debt. These, too, quickly lost value and within a year were exchanging for silver dollars at 1/24th of their face value.

Americans took to saying of the devalued paper that even the cheapest things were "not worth a Continental."

Our First Financial Czar

Most people forget that the United States had 14 Presidents before George Washington.

These, including John Hancock, held that title because they were the presiding officers of Congress under our first independent government, not the U.S. Constitution but the Articles of Confederation.

During the last years of the Revolutionary War, a wealthy Philadelphia merchant named Robert Morris became, in Rothbard's description, the "virtual economic and financial czar of the Continental Congress."

Morris pushed through a measure to make taxpayers redeem part of the national debt, debt that had seemed worthless and been sold to speculators by ex-soldiers and merchants for pennies on the dollar. This harvested a fortune for Morris's speculator friends.

Contrary to the decentralist, states' rights-oriented Articles of Confederation, wrote Rothbard, Morris pushed for creation of "a strong central government, the power of the federal government to tax, and a massive public debt fastened permanently upon the taxpayer."

Above all, Morris wanted to create America's first commercial and central bank. This Bank of North America, headed by Morris himself, would be privately owned and modeled on the Bank of England.

This new entity, chartered by Congress and opened in 1782, was America's first "fractional reserve" bank, authorized to create money by making loans based on the reserves of customer deposits...a use of other people's money that, according to Rothbard, would likely be a criminal offense in any profession other than modern banking.

Bankopoly

Bank of North America notes were by law privileged to be legal tender for paying all duties and taxes in the new nation, with value on a par with precious metal coin.

"[N]o other banks were to be permitted to operate in the country," wrote Rothbard. "In return for its monopoly license to issue paper money, the bank would graciously lend most of its newly-created money to the federal government to purchase public debt and be reimbursed by the hapless taxpayer."

"The Bank of North America was made the depository for all congressional funds," wrote Rothbard. "The first central bank in America rapidly loaned $1.2 million to the Congress, headed also by Robert Morris."

Beyond this bank's base in Philadelphia, Americans soon noticed that it appeared to be printing more paper currency than the government had metal coin to redeem.

As his political power waned, Morris repaid most government money held by his bank, ended his role as America's first central banker, and had his bank re-chartered by the State of Pennsylvania. America's first financial czar slipped quietly into history.

Seeking Sound Money

For the Framers who shaped the United States Constitution, which in 1787-89 replaced the star-crossed Articles of Confederation, their experience with inflated paper money, debt, taxation, sneaky speculators, greedy politicians and manipulative bankers left a life-long bad taste in their mouths and distrust in their hearts and minds.

For more than 100 years, popular distrust of a national bank kept America off the easy downward path that almost all other major nations have now taken.

America's winding, ascending path took us away from paper fiat currency of our Revolution and lifted us to a brief Golden Age of astonishing prosperity based on genuine, solid money.

It has been a difficult journey, however, during which the U.S. Dollar died and was brought back to life in various incarnations.

Today, in the early 21st Century the dollar has reached the last of its cat-like nine lives. It might continue as a ghost or virtual or Zombie currency. Its fast-approaching next fall, however, will end its final incarnation as tangible currency.

Constitutional Constraints

The Constitution gave Congress the enumerated power in Article I Section 8 "To coin Money, regulate the Value thereof, and of foreign Coin...." In Article I Section 10 it denied states the power to "coin money" or "make any Thing but gold and silver Coin a Tender in Payment of Debts..."

Naively, James Madison deemed this sufficient to keep American money honest. He assumed that future generations would honor the Founders' standard that government could never give itself any power not explicitly permitted by the Constitution, which gave the government no power to issue paper fiat currency.

The Constitution's Article I Section 8, alas, did give Congress the power "To borrow Money on the credit of the United States."

"I wish it were possible to obtain a single amendment to our Constitution," wrote Jefferson in 1789. "I would be willing to depend on that alone for the reduction of the administration of our government to the genuine principles of its Constitution; I mean an additional article, taking from the federal government the power of borrowing."

Piggish Bank

"Paper money has had the effect in your state that it will ever have, to ruin commerce, oppress the honest, and open the door to every species of fraud and injustice," wrote George Washington in 1787 to a constituent in Rhode Island.

The Father of our Country knew full well the threat that paper fiat money could pose to the new republic.

Yet while Jefferson served as President George Washington's Secretary of State, the President's young protege Alexander Hamilton, our first Secretary of the Treasury, eagerly planned and created the First Bank of the United States, modeled on the Bank of England.

Hamilton argued that the fledgling nation had a "scarcity" of gold and silver, and that economic growth depended on issuing paper money.

Congress established this new bank in February 1791 with a 20-year charter. The government owned 20 percent of this bank, with the rest owned by private investors.

The new bank's notes were to be redeemable in, and kept at par with, gold and silver, which in the impractical theory of a bimetallism standard were to have a fixed exchange value to one another. This gave the paper money a quasi-legal tender status.

The new money was to be invested in the assumed federal debt and in subsidy to manufacturers. Sound familiar?

"The Bank of the United States promptly fulfilled its inflationary potential by issuing millions of dollars in paper money and demand deposits, pyramiding on top of $2 million in specie [gold and silver]," wrote Rothbard.

Money Changers

"The result of the outpouring of credit and paper money by the new Bank of the United States was an inflationary rise in prices....of 72 percent," Rothbard continued. In addition, "speculation in government securities and real estate values were driven upward," along with a spawn of new commercial banks in the states.

"I believe that banking institutions are more dangerous to our liberties than standing armies," wrote President Thomas Jefferson to his Treasury Secretary Albert Gallatin in 1802.

"If the American people ever allow private banks to control the issue of their currency, first by inflation, then by deflation, the banks and corporations that will grow up around [the banks] will deprive the people of all property until their children wake up homeless on the continent their fathers conquered," Jefferson continued.

"The issuing power should be taken from the banks and restored to the people, to whom it properly belongs."

"History records that the money changers have used every form of abuse, intrigue, deceit, and violent means possible to maintain their control over governments by controlling money and its issuance," said Jefferson's protege President James Madison.

"If the debt which the banking companies owe be a blessing to anybody, it is to themselves alone, who are realizing a solid interest of eight or ten percent on it," wrote Jefferson to a friend in 1813.

Conjuring with Money

"As to the public, these companies have banished all our gold and silver medium, which, before their institution, we had without interest, which never could have perished in our hands, and would have been our salvation now in the hour of war; instead of which," wrote Jefferson, "they have given us two hundred million of froth and bubble, on which we are to pay them heavy interest, until it shall vanish into air."

"We are warranted, then," Jefferson continued, "in affirming that this parody on the principle of 'a public debt being a public blessing,'...is... ridiculous. [T]he truth is, that capital may be produced by industry, and accumulated by economy: but jugglers only will propose to create it by legerdemain tricks with paper."

"We are now taught to believe that legerdemain tricks upon paper can produce as solid wealth as hard labor in the earth," wrote Jefferson in 1816. "It is vain for common sense to urge that *nothing* can produce but *nothing*; that it is an idle dream to believe in a philosopher's stone which is to turn everything into gold, and to redeem man from the original sentence of his Maker, 'in the sweat of his brow shall he eat his bread.'"

"If Congress can do whatever in their discretion can be done by money, and will promote the general welfare, the government is no longer a limited one possessing enumerated powers, but an indefinite one...," wrote James Madison in 1792.

"Of all the contrivances for cheating the laboring classes of mankind, none has been more effective than that which deludes them with paper money," warned Daniel Webster.

The Second Bank

The first Bank of the United States lost popularity, and Congress withdrew its Charter in 1811. Soon, however, the U.S. plunged into the War of 1812, which put the young nation in what Rothbard described as "a chaotic monetary state, with banks multiplying and inflating...checked only by the varying rates of depreciation of their notes."

In 1816 Congress chartered a Second Bank of the United States. Like its predecessor, the new bank was a private corporation with one-fifth of its shares owned by the government.

This Second Bank of the United States was authorized to create paper currency, purchase a large chunk of the national debt and receive U.S. Treasury deposits.

"From its inception, the Second Bank launched a spectacular inflation of money and credit," wrote Rothbard. "Outright fraud abounded...especially at the Philadelphia and Baltimore branches....It is no accident that three-fifths of all the bank's loans were made at these two branches."

The resulting huge expansion of money and credit "impelled a full-scale inflationary boom throughout the country," wrote Rothbard. What followed was the Panic of 1819 and America's first crash and depression by 1820 as speculative bubbles in real estate and manufacturing burst.

The Second Bank survived the crisis, of which economic historian William Gouge wrote: "The Bank was saved, and the people were ruined." Again, does this today sound familiar?

The Bank War

Out of this depression's bitter fruits emerged the anti-bank movement that in 1828 elected President Andrew Jackson, a popular General who won the Battle of New Orleans against the British during the War of 1812 – and who would command a battle to destroy the Second Bank.

"The Jacksonians were libertarians, plain and simple," wrote Rothbard. "Their program and ideology were libertarian; they strongly favored free enterprise and free markets, but they just as strongly opposed special subsidies and monopoly privileges conveyed by government to business or to any other group."

"In the monetary sphere, this meant the separation of government from the banking system," he wrote, "and a shift from inflationary paper money and fractional reserve banking to pure specie [gold and silver] and banks confined to 100-percent reserves."

As President, Jackson succeeded in lowering tariffs and, wrote Rothbard, "for the first and probably the last time in American history, paying off the federal debt."

"A Den of Vipers"

"I too have been a close observer of the Bank of the United States," said Jackson in an 1834 speech. "I have had men watching you for a long time, and am convinced that you have used the funds of the bank to speculate in the breadstuffs of the country."

"When you won, you divided the profits amongst you, and when you lost, you charged it to the bank," said President Jackson. "You are a den of vipers and thieves. I have determined to rout you out, and by the Eternal I will rout you out!"

Or, as liberals put it today, the banksters pocketed the profits and socialized the losses, sticking taxpayers and investors with the bill.

Jackson withdrew government funds from the Second Bank and eventually brought it down. And when the brief economic depression of 1837 followed, the country began widespread use of Mexican silver coins as our working currency. America's economy quickly recovered.

Because of the Jacksonian movement – the free enterprise origin of what today has metamorphosed into a very different and more collectivist Democratic Party – America would do without a controlling central bank for the next eight decades.

The central bankers at last came to power in 1913 and soon got even, as we noted in our 2011 book *The Inflation Deception: Six Ways Government Tricks Us...and Seven Ways to Stop It!:*

"The Federal Reserve and the government welfare state cynically took their revenge by putting paper-money-hating Andrew Jackson's face on the $20 bill and Jefferson's face on Food Stamps to lend their legitimacy to both pieces of fiat paper that Jefferson and Jackson would repudiate were they here today."

Decentralized Banks

Without a national central bank, Americans turned to local and regional banks.

When government currency was in short supply, people bought, sold and saved by using the paper notes issued by usually-small private banks.

This was risky for savers because government regulations typically confined such banks' lending to a single state or even town, meaning that these banks could quickly run out of money and collapse if customers panicked and started withdrawing their savings. Such bank runs were all too common in the young republic.

A major cause of collapse for many such banks, however, was not customer demands for their money.

State governments "required banks to collateralize their notes by lodging specified assets (usually state government bonds) with state authorities," according to University of Georgia economist George Selgin and George Mason University economist Larry White.

"[C]lusters of 'free bank' failures were principally due to falling prices of the state bonds they held," write Selgin and White, "suggesting that the bond-collateral requirements caused bank portfolios to become overloaded with state bonds."

Or as economist Robert P. Murphy explained in *The Politically Incorrect Guide to Capitalism*, "government regulation actually unbalanced the banking system."

Government demands that banks buy large quantities of debased state bonds could push otherwise-solvent banks off a cliff, and as with today's Obama-Bernanke deliberate inflation, innocent savers paid the price for government's greed.

Dixie Dollars

Despite this, some banks developed solid reputations for safety, and their private "currency" promissory bank notes, akin to today's private bank traveler's checks, were widely accepted beyond their states' boundaries.

In the years before the War Between the States, the $10 note issued by the Citizens' Bank and Trust Company of New Orleans, chartered in 1833, may have been the most popular and trusted paper currency in many Southern states.

Because its customers included Louisiana Cajuns and many other French-speaking Americans up and down the Mississippi River, this bank note in one corner carried the large letters "DIX," French for ten.

These bank notes were quickly nicknamed "Dixies," and the Southern states where they were in wide circulation came to be called "the land of Dixies," or "Dixieland," or simply Dixie...all names that came from the regional money.

Bankrolling a War

When war between the states came in 1861, both the Union and Confederacy lacked sufficient gold and silver to pay for it. Both began issuing huge amounts of unconvertible paper money.

The Union called its paper fiat money "Greenbacks," because of their color, the same color as our fiat paper money today.

The U.S. Government insisted that Greenbacks were genuine money. It paid Union troops in this paper fiat money. Yet Greenbacks cannot even be called true legal tender because for some of its taxes – especially tariffs – the Federal Government refused to accept payment in Greenbacks; it required payment in gold.

Following the war, the victorious U.S. Government was eventually willing to redeem these Greenbacks for a small amount of gold. By then, millions had been traded to speculators here and abroad for pennies on the dollar. Those speculators became big winners when fractional gold redemption became law.

To see how the Secretary of the Treasury under Presidents Abraham Lincoln and his successor Andrew Johnson thought about gold and Greenbacks in 1865, turn to page viii of this book. You will probably never hear such economic wisdom or understanding from any modern Secretary of the Treasury.

Confederate money went the way of the paper money of America's first confederacy under the Articles of Confederation – the Continental paper used to fund the American Revolution. Many Confederate "graybacks" ended their days as wallpaper lining the rooms of homes.

Farmers and other creditors knew that Greenbacks kept being printed in huge quantities and therefore continued to lose value. This meant that Greenbacks were easy money. They could be borrowed and spent for a known purchasing value, then later repaid with cheaper, inflated Greenbacks.

By the 1870s a new Greenback Party arose, largely in the West and upper Midwest, that urged government to print lots more of this cheap paper money that benefitted debtors and shortchanged creditors.

The Greenback Party's supporters despised gold as money precisely because it did not lose value, and therefore gave them no way to repay debts or cheat lenders by using debased paper money.

The Poor Man's Gold

The Greenback Party, however, soon made common cause with Western and Midwestern populists who wanted more abundant silver, "the poor man's gold," to be the Federal Government's main coinage. The amount of silver in the economy at the time was increasing, so its supply-and-demand value was decreasing.

Bimetallism had always posed problems, as first the California Gold Rush of 1849 and later major silver strikes caused the market price of one metal or the other to vary. By 1870 it was clear that the nation should pick either silver or gold as the basis for defining the value of the U.S. Dollar.

The nation elected gold advocate candidates, and what followed were by monetary standards the most prosperous and progress-filled four decades in American history.

Speculators still caused marketplace disruptions and brief panics. Politicians still pandered for the votes of "Silverites" with laws that continued silver subsidies and coinages, and this caused investors to worry whether America would abandon gold and retreat to a silver standard.

Yet for more than 40 years Americans enjoyed the security of knowing that what they earned and saved in the form of gold was not losing value as paper fiat money always has done. During these years Americans' hard work was fully protected against government devaluation.

History has proven the French philosopher Voltaire correct: "Paper money eventually returns to its intrinsic value – Zero."

One apparent reason is that few rulers can resist the temptation to magically create out of thin air what for a time they can spend as if it were real money based on precious metal or genuine productivity.

Over the Rainbow

As we noted in our 2010 book *Crashing the Dollar: How to Survive a Global Currency Collapse*, America's 19th Century political struggle between gold and silver may have led to Oz.

Some scholars believe that L. Frank Baum's 1900 tale *The Wonderful Wizard of Oz* is actually a parable about monetary reform in which the sure road is gold, Dorothy's slippers were in the original book silver instead of ruby, and Oz is the abbreviation for Ounce, the measure of gold and silver.

The story reflects the great debate of the 1890s in the upper Midwest, these scholars say, over whether government should issue gold-backed or silver-backed currency. This debate gave us 1896 Democratic presidential candidate William Jennings Bryan's famous "You shall not crucify mankind upon a cross of gold" speech.

Baum at the time was editor of a South Dakota newspaper in the midst of such populist activism, and his book was published in 1900, the year of Democrat Bryan's second unsuccessful presidential run.

Americans chose to re-elect Republican President William McKinley, who in 1900 had signed into law the Gold Standard Act, to continue an era of growing prosperity for most Americans.

Rough Rider

After McKinley's 1901 assassination, Americans suddenly found that the reins of government had been passed to his 42-year-old Vice President Teddy Roosevelt, who had been the aggressive, Progressive reformist Governor of New York and media-lionized national hero who led a combat group known as the Rough Riders in Cuba during the Spanish-American War.

This first President Roosevelt would set forces in motion that led to the end of America's Golden Age – the genesis of the continuing downhill slide of the dollar's value, of 100 years of the U.S. Dollar's Great Debasement, and our current age of "Modern Money."

Soon the value of the U.S. Dollar would be put under the control of the monetary wizards at a mysterious and secretive, all-powerful central bank that now shapes our world.

These Progressives would crucify humankind not upon a cross of gold, as Bryan said, but upon a double-cross of paper.

They soon would replace America's good-as-gold dollar with a relentlessly-debased, devalued paper fiat currency that would subject America's economic strength, stability, security and prosperity to death by a thousand paper cuts.

"The great majority of central banks were established after 1900
to help governments spend money they didn't have.
They became engines of inflation.
The largest number of runaway inflations and the worst
runaway inflations have occurred since 1900."

– Jim Powell
Forbes Magazine
November 29, 2011

"Like every other tax, inflation acts to determine
the individual and business policies we are all forced to follow.
It discourages all prudence and thrift.
It encourages squandering, gambling, reckless waste of all kinds.
It often makes it more profitable to speculate than to produce.
It tears apart the whole fabric of stable economic relationships.
Its inexcusable injustices drive men toward desperate remedies.
It plants the seeds of fascism and communism.
It leads men to demand totalitarian controls.
It ends invariably in bitter disillusion and collapse."

– Henry Hazlitt, economist

"The price level in 1929 was not much different...
from what it had been in 1800. But, in the two decades
following the abandonment of the gold standard in 1933,
the consumer price index in the United States nearly doubled.
And, in the four decades after that, prices quintupled.
Monetary policy, unleashed from the constraint of domestic gold
convertibility, had allowed a persistent overissuance of money
....fiat currency was inherently subject to excess."

– Alan Greenspan
Chairman, Federal Reserve
December 19, 2002

Chapter Three
The Progressive Takeover

"The statesman who should attempt
to direct private people in what manner
they ought to employ their capitals,
would...assume an authority
which could safely be trusted to no council
and senate whatever, and which would
nowhere be so dangerous as in the hands of
a man who had folly and presumption enough
to fancy himself fit to exercise it."

– Adam Smith

In the years leading up to 1913, America was undergoing a huge transformation that changed both of our major political parties, our government, our economy and the fate of the U.S. Dollar forever.

Thirty-two million of America's 92 million people, according to the 1910 Census, lived and worked on farms. Their way of life was being replaced by agricultural machines so rapidly that within a century America would be producing four times more food than we consume with only two percent of Americans farming. Fifty-four percent of Americans lived in communities of 1,000 or fewer people.

People from rural farms and foreign lands were flocking in pursuit of happiness to factory towns and big cities. Most found jobs and a rising standard of living, yet they also found cities run by corrupt political machines, and industries owned by the corporations and Trusts of the wealthy and powerful.

The spirit of the age was "progress" and change. Scientists and inventors were remaking the world with electric light, gasoline-powered vehicles, radio communications, phonographs and flying machines.

In this era of upheaval, from Russia to rural America, many well-intentioned people began asking if government and society could be improved. Many wondered if the old ruling elites should remain in control of this new world.

"Progressivism" became the umbrella label that brought together populists suspicious of private or public concentrated power and wealth, a variety of social reformers, and recent immigrants, some of whom came to America bringing European socialist ideas of class warfare and expropriating the rich.

"This Invisible Government"

In 1912 former Republican President Teddy Roosevelt, after being denied his party's nomination, sought re-election to the White House on his own Progressive Party ticket.

His "Bull Moose Party" platform advocated women's suffrage (unlike either of his rivals, Progressive Democrat Woodrow Wilson and Republican incumbent William Howard Taft). It also called for a National Health Service; social insurance for the disabled, elderly and unemployed; an eight-hour workday; an inheritance tax; and a constitutional amendment permitting an income tax, among many other things.

Above all, Roosevelt's Progressivism centered on his "New Nationalism" – his ideal of a paternalistic, muscular central government to regulate businesses, create great national projects such as the Panama Canal and national parks, extend American values to other lands, and protect working people and the middle class.

"To destroy this invisible Government, to dissolve the unholy alliance between corrupt business and corrupt politics is the first task of the statesmanship of the day," declared Roosevelt's 1912 platform.

Power-loving Progressivism

Democratic Progressive Woodrow Wilson won the presidency with 42 percent of the national vote. Roosevelt split the Republican vote, winning 27 percent of the popular vote to his one-time friend President Taft's 23 percent. Socialist Party candidate Eugene V. Debs won 6 percent.

The following year, 1913, the Great Debasement began. President Wilson would transform America forever by signing an income tax law made constitutional by the new 16th Amendment. He also signed the act creating the Federal Reserve System, the quasi-private entity that has grown to control America's banking institutions, money supply, economy and more in fulfillment of an evolving Progressive ideology.

Wilson signed into law the Clayton Antitrust Act to regulate corporations and trusts, and the Federal Trade Commission Act giving government far-ranging powers to intervene in business.

Most remember Woodrow Wilson, who was President of Princeton University, as America's only professional intellectual President. (Jefferson was a far greater intellect, but not a professional academic.)

Most also remember that Wilson took the United States into World War I only months after being re-elected in November 1916 under the slogan "He Kept Us Out of War."

After that war, Wilson and his "Fourteen Points" helped establish the League of Nations, forerunner of today's United Nations; the U.S. Congress, unwilling to relinquish a measure of America's sovereignty to the international body, refused to ratify Wilson's agreement to join the League.

Wilson was fashionably Progressive, pro-organized labor and anti-Big Business. Yet his smug Progressivism was also eager to impose his eccentric morals on others.

Wilson cracked down on the free speech of those who opposed his entry into the war. Recreating his own version of the Alien and Sedition Acts that Thomas Jefferson had abolished, Wilson unleashed his Attorney General A. Mitchell Palmer, who rounded up 10,000 reds and other radicals and deported many of them.

Wilson, born in Virginia and raised in Georgia, was the first Southerner to be President since 1869. His father was a slaveowner and Confederate chaplain. One of Wilson's earliest boyhood memories was of standing next to Robert E. Lee and looking up at the Confederate general's face.

As a Progressive president, Wilson re-segregated the federal civil service that Republicans had racially integrated. African-American federal employees were required to eat in segregated dining rooms and use "Blacks Only" bathrooms.

Wilson also enthusiastically praised and encouraged Americans to see the D.W. Griffith movie "Birth of a Nation," which was openly pro-Ku Klux Klan.

Yet Wilson's racism did not necessarily extend to other racial or ethnic groups. While President, after his first wife died, Wilson wed a direct descendant of famed Native American princess Pocahontas.

Wilson's Democratic Party has always used racial polarization politics to gain and hold power. What is amazing is that those who have historically been this party's most abused victims are today its most loyal voters.

The Second Progressivism

In 1920, following World War I, voters restored Republicans Warren G. Harding and then Calvin Coolidge to the White House. The war's economic undertow pulled America into what could have become the Great Depression.

Instead of responding as Progressive Franklin D. Roosevelt did 12 years later, with massive government programs and stimulus spending, Harding and Coolidge simply got government out of the way. An unhindered free market quickly adjusted, and the economic downturn turned into the prosperity and good times of the Roaring '20s Jazz Age in less than two years.

In 1924 a second Progressive Party emerged, centered on another Republican, Robert LaFollette, the energetic reformist Senator of Wisconsin.

Where Teddy Roosevelt wanted to regulate giant corporations, "Fighting Bob" LaFollette hated them and sought to stamp them out. He had also been an outspoken critic of America's involvement in World War I.

This second wave of Progressivism came after the Russian Revolution. Many Progressives felt energized by the revolutionary idea that old social orders, including capitalism, could be swept aside and replaced with something that claimed to be better. Most would later become deeply disillusioned as they recognized the real nature of the Soviet Union.

Where Teddy Roosevelt's Progressive Party was largely made up of White Anglo-Saxon Protestant (WASP) social reformers and was critical of those who considered themselves "hyphenated Americans," LaFollette's Progressive Party was much more oriented to blue-collar organized labor, and to racial and ethnic minorities as well as recent immigrant groups who very much still thought of themselves as hyphenated.

Part of LaFollette's vision of something better was a future that put academic social planners in control. To this end, he encouraged the University of Wisconsin-Madison to play a significant role advising the state government.

In the 1924 campaign, the charismatic, emotional LaFollette was endorsed by the American Federation of Labor (AFL) and the Socialist Party of America.

He got 17 percent of the popular vote nationwide, yet won only the 13 Electoral Votes of Wisconsin, where his sons would later create a small dynasty of state officeholders. LaFollette finished second in 11 Western states. After this election his Progressive Party was finished and disbanded.

Third Wave Progressives

Following World War II, in 1948 a third Progressive Party arose to challenge President Harry Truman. This party's standard-bearer was Henry Wallace, who had been FDR's Vice President before Truman.

Less than six months after FDR replaced Wallace with Truman, President Roosevelt died. Truman became President.

Henry Wallace was by most accounts not a Communist, yet he demonstrated affection and admiration for Josef Stalin's Soviet Union, which had been America's ally during World War II. For four years, this Soviet sympathizer was a heartbeat way from becoming President while FDR's health was in steep decline.

The 1948 Progressive Party, however, by most accounts was heavily infiltrated and influenced by Communist Party apparatchiks and ideologues. Socialist Norman Thomas and several other prominent leftists resigned from the party over what they called undue Communist influence on Wallace.

Every other page of the 1948 Progressive Party Platform blames the United States for the dawning Cold War, and urges friendly relations with the Soviet Union, whose army was then occupying Eastern Europe and setting up puppet regimes.

The problems in America, this 1948 platform said, were caused by "Big Business control of our economy and government.... Today that private power has constituted itself an invisible government which pulls the strings of its puppet Republican and Democratic parties.."

These "old parties," the Progressive Platform said, "refuse to negotiate a settlement of differences with the Soviet Union.... They use the Marshall Plan to rebuild Nazi Germany as a war base and to subjugate the economies of other European countries to American Big Business.... They move to outlaw the Communist Party as a decisive step in their assault on the democratic rights of labor, of national, racial, and political minorities, and of all those who oppose their drive to war.... We denounce anti-Soviet hysteria as a mask for monopoly, militarism, and reaction...."

The 1948 Progressive Party platform goes on like this for more than 30 pages, calling for the government to seize ownership of the means of production from private corporations, denouncing all anti-Communist investigations by Congress, and so forth.

In the presidential election, Henry Wallace won 2.4 percent of the popular vote and zero Electoral Votes. Dixiecrat Strom Thurmond also won 2.4 percent of the national vote, but he carried South Carolina, Alabama, Mississippi and Louisiana plus one electoral voter in Tennessee for a total of 39 Electoral Votes.

Today's Progressives

Many of today's Fourth Wave Progressives developed their political ideas from the anti-Vietnam War, Civil Rights and New Left movements of the 1960s and 1970s. Their views resonate with the Social Democratic political parties of Europe, which describe themselves as democratic socialist in ideology.

The 76 members of the Congressional Progressive Caucus are almost all Democrats. In past years this caucus openly embraced the group Democratic Socialists of America (DSA), of which labor leaders such as AFL-CIO President John Sweeney were proud card-carrying members.

Beginning with President Ronald Reagan's run for re-election in 1984, the Communist Party USA has ceased running its own candidates and directed its followers to cast their vote for the Democratic Party candidate. The CPUSA does this to avoid splitting the coalition of the political left.

The Democratic Party certainly does not seek CPUSA support, nor does such support suggest in any way that Democratic candidates share Communist views.

This might mean, however, that America's Communist Party is either terrified of the Republican Party since Ronald Reagan, or that today's Progressive-dominated Democratic Party has shifted far enough to the left that Communists find its policies at least minimally tolerable.

In 2007 then-Senator Hillary Clinton during a presidential debate was asked if she described herself as a liberal. "No," replied Ms. Clinton, "I consider myself a proud modern American progressive, and I think that's the kind of philosophy and practice that we need to bring back to American politics."

"Progressive" has become a fashionable designer label to wear among Democrats nowadays, in part because it is remarkably vague.

Judging by its three previous waves of evolution, the label called Progressive suggests that its wearer's thinking might be at least one part Progress, one part anti-Big Business populism, one part democratic socialism, one part class warfare, one part smug Nanny Statist "We Know Best" do-goodism, and one part liberalism.

Procrustean Progressives

So who are today's Progressives, and what do they believe?

The chief political values of traditional Americans have always been liberty, rugged individualism, and taking responsibility for one's own life, family and values.

The foremost ideological values of Progressives in some ways seem to come more from the French Revolution than the American Revolution. Progressives primarily desire equality and fraternity, i.e., a belief that the collective, the group however defined at any given moment, matters more than the individual.

In a Progressive world, everyone is equally special – which means that nobody is special. Exceptional qualities, ideas and achievements are weeds in the Progressive garden of Egalitarian Eden, where individuality is to be torn out by the roots and removed.

By equality, Progressives a century ago and today do not primarily mean equal treatment and rights under the law. They seek to use government power to end the inequality of rich and poor, man and woman, gay and straight, black and white, and much more.

Yet, paradoxically, Progressives have always been eager to have government treat people unequally in order to impose their egalitarianism. They are delighted to confiscate wealth from the successful so government can redistribute these riches.

The successful, as Progressives like to say, are merely "winners in life's lottery" – a phrase that disparages the successful as merely lucky, not meritorious. As to the poor, in a folk song made popular by singer Joan Baez, "there but for fortune go you and I."

Progressives likewise fight racism by using reverse racism – laws and government policies that discriminate in favor of members of one race over others in matters of preferential college admissions, hiring, quotas and other opportunities.

To achieve egalitarian "leveling" of society, Progressives have always been eager to make government heavier and heavier in order to hammer down whatever individualistic nails stand out from the masses.

To fight business "monopolies," for example, Teddy Roosevelt and Woodrow Wilson paradoxically were happy to expand the power of the biggest monopoly of all, the government.

Ancient Greeks told of the fabled character Procrustes, who to fulfill his notions of equality would waylay travelers and tie them to his Procrustean bed.

Those Procrustes deemed too short for this bed would be painfully stretched to make them longer. Those who were too tall would have their feet, legs or even head cut off to shorten them to the bed's length, in conformity with Procrustes' notion of equality.

"Equal" Yet Superior

Progressives claim to support the ideal that all people are equal. Yet at the same time Progressives regard themselves as morally and mentally superior to those who hold different values.

As we have learned during the past 100 years, were it not for their double standards, Progressives would have no standards at all.

The founding father of psychiatry, Sigmund Freud, identified a mental tendency he called "projection" that makes people project or ascribe what they are like onto others. A thief, for example, will tend to assume that other people are thieves.

Progressives tend to assume that all conservatives, especially Christians, are "trying to impose their morality on everybody."

Nearly all Progressives believe that their superior morality and intelligence entitles them to impose their values onto everyone else through the force of government.

The 18th Amendment to the Constitution – Prohibition of alcohol – was enacted in 1919 during Wilson's second term by a bipartisan coalition of well-meaning religious conservatives and Progressive social activists.

Paternalistic Progressives

This Progressive mindset could be heard in the 1984 Democratic Convention speech of then-New York Governor Mario Cuomo. Boiled down to its essence, Cuomo's message was that we are all one big family, and that government is the parent who should decide how much each family member must pay into this collective, and how much each will receive from it.

In Cuomo's paternalistic Progressive society, government is the master parent, not the public servant, and we are children who must accept their inferior place and bow to its superior Progressive wisdom.

Political science has a formal term for the kind of social-political system Governor Cuomo and other Progressives advocate. That term is feudalism, the system of rulers, castles and subservient serfs of the dark and medieval ages.

The biggest difference between then and now is that feudal serfs typically were forced to pay only 10 percent of everything they produced to the lord of the local castle.

In today's Progressive neo-feudalism, average Americans are forced to pay federal, state, local and hidden taxes of around 55 percent of what they earn to our Progressive governments.

In medieval England when the rulers King John and his agent the Sheriff of Nottingham turned rapacious, legend tells that a hero we remember as Robin Hood seized money the government had stolen – and returned it to its rightful owners, the taxpayers.

Back then, a starving peasant would be put to death for killing and eating a wild deer in Sherwood Forest – because these deer, like everything else in nature, were defined by law as the King's property.

Eco-ideology

Today's radical environmentalist Progressives have gone even farther in their craving for power. Even after a Democrat-controlled U.S. Senate voted against his measure to give government total control over the "greenhouse gas" carbon dioxide as a "hazardous pollutant," President

Barack Obama had his Environmental Protection Agency assert that it had a regulatory authority to control all sources of this gas.

As you read this, you are naturally exhaling carbon dioxide, a by-product of human breathing. President Obama therefore has asserted his Administration's power to directly regulate and control you and all other carbon-based life forms that breathe.

Progressives use both taxation and regulation to force their values onto others. Do not be surprised if you soon must pay a breathing tax. President Obama opened the door, in effect, to changing the traditional Presidential song "Hail to the Chief" to "Exhale to the Chief."

Progressive nanny statism also continues today in New York City Mayor Michael Bloomberg, who has imposed government restrictions on every individual's access to salt, sugar and foods containing trans-fats – not only for children in public schools, but also for adults in private restaurants.

To the left of Mayor Bloomberg are even more extreme Progressives who believe that those who disagree with them on issues such as global warming should be fired, denied media access, and in rare cases turned into targets for violence. Procrustes would feel a kinship with such Politically-Correct Progressives.

A Genuinely "Progressive" Income Tax

In 1913 the 16th Amendment became part of the U.S. Constitution. It permitted creation of an income tax. The Federal income tax law enacted immediately thereafter was "Progressive," imposing heavier tax on those with larger incomes.

The income tax was passed by Congress amid promises that it would tax "the rich." This technique, appealing to envy, has been used ever since to enact nearly every tax expansion, e.g., the Alternative Minimum Tax (AMT). Beginning in 2013, AMT is scheduled to increase taxes for at least 34 million Americans, most of whom are not wealthy.

In 1913, when one lawmaker proposed putting a 7 percent tax ceiling into the income tax law, colleagues ridiculed him by saying "If we did that, then some fool would try to raise the tax that high."

The 1913 income tax imposed a marginal tax rate of 1 percent on the first $20,000 of income for married joint filing taxpayers, according to the Tax Foundation. That tax rose to 2 percent on incomes up to $50,000; 3 percent on up to $75,000; 4 percent on up to $100,000; 5 percent up to $250,000; 6 percent on up to $500,000; and 7 percent on any income above $500,000.

We need to remember that the 1913 dollar had not yet undergone 100 years of the Great Debasement, as our dollar has.

Remember, too, that in 1913 the average American earned $740 a year.

A one-percent income tax on this 1913 American, even with zero deductions, would have been $7.40, less than three days' income each year.

That average American's $7.40 income tax in 1913 would today, if increased by inflation alone, be somewhere between $151 and $370. Sound good?

If you pay vastly more than this today, it is because the Progressives' Great Debasement has destroyed the value of your dollars.

Your tax in 1913 would not jump to 2 percent until you earned $20,000 – the equivalent in 2012 of earning $453,292.

To reach 1913's top marginal tax rate of 7 percent, you needed to earn $500,000, which – as calculated by inflation alone by the Tax Foundation – in 2012 dollars is $11,332,304.

Progressive and neo-Marxist economists dismiss such comparisons by arguing that wage inflation has kept up with price inflation. This argument is misleading for a host of reasons.

We need to keep at least three realities in mind:

(1) The Progressives have used the very inflation their Great Debasement created to force average Americans into confiscatory tax brackets once used to tax only the wealthy.

The issue is no longer how much you "earn," but how much the government lets you keep for yourself. Under their divide-and-conquer politics, half the working-age population now pays no income taxes. Those who do pay carry a much heavier-than-equal share of the burden.

(2) If income has kept up with inflation, then why must both husband and wife work in today's average American household – when the income of one earner used to be enough?

Many millions with fixed or limited incomes have not been able to keep up with the Progressives' deliberate inflation. An even more grim reality is that in most families one spouse is working just to pay the taxes and other costs, visible and invisible, now imposed by Progressive government.

(3) Back in 1913, the gold standard was still in effect. With its fixed exchange rate around $20 per ounce, a person could readily convert $500,000 in U.S. Dollars into 25,000 ounces of gold – or 25,000 $20 gold coins.

At gold's September 2012 price of around $1,775 per ounce (and adding no numismatic value for what in 1913 were everyday, circulating coins), $500,000 in such 1913 gold coins would in September 2012 be worth approximately $44,375,000 – nearly $44.4 Million when calculated in gold.

The average American has always earned less than this, of course, but in 1913 any savings could easily be done in the safe, value-growing medium of gold numismatic coins, beyond the reach of the Great Debasement.

From 7 to 94 percent

The new income tax increased drastically in 1917, the year the United States entered World War I. Tax brackets multiplied dramatically. Suddenly married joint filers were required to pay 2 percent on income up to $2,000 ($35,059 in 2012 dollars, adjusted for inflation), and 4 percent on earnings to $5,000 ($87,648 in 2012 dollars).

The top marginal tax rate in 1917 had soared to 67 percent on all income above $2 Million (just over $35 Million in 2012 dollars).

The greed of Big Government Progressives has been insatiable. By 1936 they pushed the top marginal income tax rate to 79 percent, then to 94 percent in 1944, backing off to "only" 91 percent from 1946 until the mid-1950s.

The top rate was still 70 percent when President Ronald Reagan was inaugurated in 1981; in two terms he drove the income tax top rate down in stages to 28 percent in 1988, producing a huge, long-lasting economic environment of growth and prosperity.

When Progressives regained power under President Bill Clinton, they rushed to jack the top tax rate back up by more than 41 percent – to 39.6 percent in 1993.

Progressives screamed when Republican President George W. Bush in 2003 lowered the top tax rate to 35 percent – still 25 percent more than it had been under President Reagan, and still taking more than a third of the income of those it hit.

President Barack Obama vowed that if re-elected in 2012, he would greatly increase tax rates on payers in all income tax brackets – and would slash a wide range of deductions and tax credits to push government's take from taxpayers even higher.

Slamming Shut the Gold Window

During World War I, Progressive President Woodrow Wilson had effectively restricted the gold standard. This made it more difficult to convert paper gold certificates into gold itself, and immensely harder to take monetary gold out of the United States without the approving signature of one or more high government officials.

People suddenly found that their escape hatch from debased fiat dollars into the secure safe haven of gold had been slammed shut and locked, at least until after World War I. Within 13 years, that door would be closed, and remain locked for more than four decades.

American lawmakers discussed creating an income tax during the War of 1812 and actually imposed it during the War Between the States.

Democrats enacted an income tax in 1894 that imposed a 2 percent tax on all incomes above $4,000 – which meant that the tax would hit roughly the top 10 percent of households in America. The U.S. Supreme Court struck the tax down because the Framers had in the Constitution wisely prohibited any tax that targeted individuals by requiring that all taxes be apportioned among the states.

A Hidden Agenda

Prior to the Progressives, even those who supported an income tax generally did so to fund the government – during wars in 1812 and 1861, or to offset revenue lost from reducing a tariff in 1894.

In 1848 in *The Communist Manifesto*, Karl Marx and Friedrich Engels proposed 10 steps to destroy the capitalist bourgeoisie, the class of business people and the wealthy. Nine of these 10 steps have already, either entirely or in part, been implemented in the United States.

Three of Marx's and Engels' steps:

1. Restrictions should be put on inheritance of wealth, at least by "rebels" who oppose the state.

2. All banking and credit should be centralized under government control.

3. Government should impose a heavily-graduated "Progressive" income tax that confiscates much more money from the rich than from others.

Today's Progressives are quick to say they are not Marxists. Many, however, share Karl Marx's hatred for the wealthy and for prosperous business people, and for some of the same reasons.

Most Progressives, including President Barack Obama, openly favor government redistribution of wealth.

In Mr. Obama's phrase to the citizen who came to be known as Joe the Plumber: "Things just work better when we spread the wealth around."

Note the planted assumption here that it is "*the* wealth," not *your* wealth, not the property of someone who earned it and rightfully owns it. Wealth in Barack Obama's phrase is a shared entity, the way one would speak of "the sky" or "the ocean."

If some are needy, the Progressive assumption goes, then it is not only ethically acceptable but also morally imperative to take the "excess wealth" from those not in need and give it to the poor.

From a conservative and libertarian view, this Progressive ideological point of view poses many ethical problems. This redistribution of wealth is coercive.

It puts the coercer in the morally-dubious role of deciding whose need warrants such expropriation of wealth from one and redistribution to another. The person arrogant enough to do this, as Adam Smith noted, is precisely the person least fit to do so wisely or justly.

To Eradicate the Rich

Let's acknowledge an unpleasant truth about Progressives, who tell us they care about the poor.

If tomorrow, America's wealthy established an immense trust fund that guaranteed every poor person in the country food, shelter, clothing, education, medical care and every other necessity of a decent life, would Progressives be content?

We all know the answer. No, Progressives would continue to hate the rich and continue to press for confiscatory taxes on them.

Progressivism does not want to raise the poor up. Progressivism wants to tear the rich down.

Progressivism needs to keep the "poor" poor in order to justify its many expansions of government in their name, including a steeply-graduated income tax of the sort Marx wanted to impose on the rich. To Progressives, the poor are a means to an ideological objective.

Under President Lyndon Johnson's War on Poverty, approximately 80 cents of every dollar spent for the poor never reached the poor. The bulk of this money went to support an array of social workers, "poverty pimps," community organizers like the young Barack Obama, $100,000-a-year political appointees and others who justify their own substantial income and power in positions spending this government money.

In today's War on Poverty, the government in 2012 will spend an average of $62,000 on every family in poverty. As the late Nobel laureate economist Milton Friedman proposed, why not give this – or some large fraction thereof – to the poor family directly? If government did this, every poor family's income would exceed the national average. Progressives will not accept this, however, because it would eliminate the bureaucratic and politician middlemen in the deal. It would no longer expand and empower the government, which was the real purpose of this redistributionist program.

Independent Islands

Why do Progressives hate the rich?

The rich are like little independent islands, with their own power and freedom and individual economic, intellectual and spiritual sovereignty in what Progressives want to be a monolithic sea that Progressives control and where no dissent from Progressive Politically Correct orthodoxy is tolerated.

The rich on their independent islands can provide refuge for those who do not want to bow to Progressivism.

The Koch Brothers for many years funded the libertarian think tank the Cato Institute that advances alternatives to Progressive Big Government ideology.

Rupert Murdoch funds the Fox News and Fox Business cable channels as well as the *Wall Street Journal*, *New York Post* and other outposts of free thought.

Strip the wealthy of their money, as Progressives are constantly striving to do, and this would eliminate almost every bastion of non-Progressive free thought.

With the wealthy gone, all that would then remain is an all-powerful government, government-funded leftist universities, and today's sycophantic and partisan Progressive mainstream media.

Progressive Judgment

Progressives tend to see business and the profit motive as evil. They imagine that business people sit around all day counting their money, chanting "Greed is Good," and cheating customers.

The reality, as 18th Century British author Samuel Johnson observed, is that "A man is seldom more innocently occupied than when he is engaged in making money."

The capitalist succeeds or fails by risking his own investment money in a realm of voluntary transactions. No one is forced to buy his product or service. Those who believe he cheated them will never be his customers a second time, so he has every incentive to deal honestly with them.

To succeed in the marketplace, the business person must strive to provide the best product and service at the best price to customers who "are always right" and can take their business elsewhere. He makes his income by peaceful persuasion, without resort to force or coercion.

Progressives say they favor freedom and choice, yet they want government to interpose itself between people who freely choose to swap goods, services and money with one another. Progressive freedom apparently does not extend to this form of human interaction that we call enterprise.

Compare this to what Progressives deem superior to capitalism: government.

"Government is not reason, it is not eloquence – it is force," goes a statement widely attributed to George Washington. "Like fire, it is a dangerous servant and a fearful master."

Government does use coercion, force and the threat of force, to take whatever it wants from its subjects – their liberty, property, earnings – and to redistribute what it takes to the ruler's cronies and favorites.

Government makes almost nothing – except war. What it spends is taken, in one way or another, from others.

As we have said before, Progressives such as Teddy Roosevelt accused capitalists of forming "monopolies." (In a free marketplace, monopolies are almost impossible to sustain without the assistance of government's coercive power.)

Yet today's Progressives are eager to give total power to the ultimate monopoly and the wealthiest, greediest, most coercive entity in society – the government.

A Wealth Tax

In 2012 the Progressive income tax has become not a tax on income but a tax on the upper-middle and upper classes. Those with the top 10 percent of incomes now pay more than 70 percent of all income taxes. Progressives such as President Obama obsessively demand that taxes on "millionaires and billionaires" be raised even more to force the successful to "pay their fair share."

One of us once during a radio interview with Progressive media star Alan Colmes asked: is there any level of taxation – 90 percent? 95 percent? 98 percent? – at which he would agree that a wealthy person was paying enough and should be subjected to no more tax increases?

Colmes, like many of today's Progressives, refused to agree to any tax ceiling for the rich, no matter how high. The apparent reason: Progressives want it all. They want the rich to cease to exist...except, of course, for Progressive plutocrats. As we have said before, if not for their double standards, Progressives would have no standards at all.

This new tax that Progressives pushed into law in 1913 has always violated the Equal Protection principle of the Constitution by imposing no burden on millions of Americans while initially imposing a 7 percent – and today a vastly heavier – burden on other Americans.

This Progressive tax also violated American values in other ways. It penalized success and disproportionately targeted the successful for audits. And it rewarded failure with a far lower rate of taxation.

Truly "Progressive" Taxation

A genuinely "Progressive" income tax – one that encouraged progress and prosperity for all – would either impose no added burden on success whatsoever, or at least would have tax rates that became progressively lower as a person's income increased, thereby giving people an incentive to work harder to lower their taxes by earning more money.

Millions would do that occasional extra hour of overtime if it bumped them into a lower tax bracket instead of a higher one.

The Progressive income tax is un-American in many ways almost too self-evident to mention. To name just one, it implicitly gives government a legal rationale for doing surveillance to determine how much money Americans earn.

Prior to the income tax, a citizen's personal income was mostly regarded as none of the government's business. The income tax suddenly gave tax agents the authority to pry into every transaction, and to cut government in for a potential piece of every person's profits.

The very Progressive American Civil Liberties Union has defended citizens and groups when government agents have done as little as put a few newspaper clippings about their activities into a file. Why has the ACLU never challenged the legal power of the Internal Revenue Service to require and to keep detailed personal financial records on more than 100 million Americans?

Orwellian Progresspeak

Progressives have developed a special jargon to advance and obscure their agenda. This jargon has much in common with Orwellian doublethink, as used in George Orwell's dystopian novel *1984*.

In Orwellian Progresspeak, letting someone keep more of what they earn in a tax cut is a government "expenditure" that must be offset by raising their, or somebody else's, taxes.

In this mind-warping jargon of Progressives, government spending is an "investment" – even though it is unlike any free market investment because it has zero prospect of earning any profit.

Having the audacity to call government spending "investment," however, invests the tawdry funneling of taxpayer money to political cronies with an aura of collectivist superiority.

Roads and other infrastructure may, indeed, enhance commerce, yet they are no substitute for voluntary investment in the free marketplace.

Progressives, motivated by politics and an ideology of redistributing wealth, are more likely to build a bridge to nowhere with taxpayer money than a bridge of optimal social utility and efficiency.

They also create an unhealthy collectivist mindset in some radical Progressives that "if you have a successful business, you didn't build that," that all success must be credited to, and hence belongs to, the collective and the state.

As *New York Times* Progressive columnist and economist Paul Krugman has written, "Government is not a business," and therefore it need not balance its books or justify its income and outgo like a business.

Krugman thus dismisses critics of big spending who note that any business person who was caught playing the phony bookkeeping games our government uses every day would be in prison. Krugman is correct, however, that we should never think of government as an "investor" in the noble sense we would ascribe to businesses.

Progressives made a major effort through the McCain-Feingold Campaign Finance Law to restrict or prohibit campaign ad funding by corporations on the grounds that "a corporation is not a person." The U.S. Supreme Court struck down this blatant attempt to stifle the views of companies.

If corporations are not persons, we ask, then how does government justify taxing them? Was not a founding idea of the American Revolution "No taxation without Representation"? We do not allow corporations to vote in our elections. Progressives tried to silence all corporate political free speech during election campaigns. How, then, can a corporation have representation that legitimizes making it pay taxes? Just a thought inspired by the Orwellian surrealism of Progresspeak, the speech and thought of Progressives.

*"Like gold, U.S. Dollars have value only
to the extent that they are strictly limited in supply.*

*"But the U.S. Government has a technology, called a
printing press (or, today, its electronic equivalent),
that allows it to produce as many U.S. Dollars
as it wishes at essentially no cost.*

*"By increasing the number of U.S. Dollars in circulation,
or even by credibly threatening to do so, the
U.S. Government can also reduce the value of a dollar
in terms of goods and services, which is equivalent
to raising the prices in dollars of those goods and services.*

*"We conclude that, under a paper-money system,
a determined government can always generate
higher spending and hence positive inflation."*

– Ben S. Bernanke
Member, Board of Governors
Federal Reserve Board
(and now Fed Chairman)
November 21, 2002

Chapter Four
The Great Debasement

"Whenever destroyers appear among men, they start by destroying money, for money is men's protection and the base of a moral existence. Destroyers seize gold and leave to its owners a counterfeit pile of paper. This kills all objective standards and delivers men into the arbitrary power of an arbitrary setter of values."

– Ayn Rand, *Atlas Shrugged*

"The Great Recession shook America to its foundations in 2008 and 2009. And to tell the truth, it's never ended," said Professor Pat. "We're still deep in this recession."

"The Great Depression that began in 1929 lasted for more than 10 years, we're taught," he continued, "but I suspect that it never really ended either. As the last great nation standing, with our banks and factories undamaged, we came out of World War II with vast amounts of power and credit – and we used both to create what looked like prosperity. Yet even from the 1940s and '50s until today, we've been living on a credit card and in an illusion."

Ryan and Peggy looked puzzled. "It sure felt like prosperity to us," said

Ryan.

"Everything Else is Credit"

"So does living it up on a credit card, until the bill arrives," Patrick said. "For America the 1940s and '50s were the height of American power. We had become an empire with global reach, and the bills for this would not arrive until the 1960s and '70s.

"It's what the tycoon J.P. Morgan told Congress in 1912. 'Gold is money,' said Morgan. 'Everything else is credit.'"

In 1933 President Franklin D. Roosevelt made it illegal for most Americans to use gold as money. FDR killed the gold-backed dollar, except for specific U.S. Government and Federal Reserve dealings with certain foreign central banks.

In other words, in America we have not been using genuine money since 1933. We have merely been using paper fiat currency, which in the pure sense is not actually money at all.

"Everything we Americans have thought of as money and prosperity, from 1933 until today, has really been credit," Professor Pat continued. "That's all that paper fiat money, with no intrinsic value of its own, ultimately is – a promissory note, a kind of credit coupon whose value declines as government prints more."

Running Up the Tab

Here's how Ayn Rand explained this in her novel Atlas Shrugged:

"Gold was an objective value, an equivalent of wealth produced.
Paper is a mortgage on wealth that does not exist,
backed by a gun aimed at those who are expected to produce it.
Paper is a check drawn by legal looters
upon an account which is not theirs:
upon the virtue of the victims.
Watch for the day when it bounces,
marked, 'Account overdrawn.'"

This prosperity built on credit and debt began to crumble after President Richard Nixon in August 1971 ended the last tenuous anchor tying gold to what were popularly called Eurodollars. These were dollars held as reserves by several European central banks that those banks were allowed to redeem for gold at $35 to the ounce.

When Mr. Nixon cut the U.S. Dollar adrift, its value rapidly fell by about one-third. Within a decade the dollar had lost half its 1971 value.

The result was soaring inflation, sky-high mortgage rates, and oil embargoes by suppliers angry at being paid in debased dollars.

Overdrawn

As financial analyst Egon von Greyerz of the Swiss firm Matterhorn Asset Management describes it, the U.S. Consumer Price Index (CPI) had been relatively stable from America's founding until the early 1900s. From Nixon's de-gilding of the dollar in 1971 until 2010, however, the CPI jumped by more than 500 percent.

"The reason for this," von Greyerz says, "is uncontrolled credit creation and money printing."

"Total U.S. debt went from $9 Trillion in 1971 to $59 Trillion [in 2010]. U.S. nominal GDP (Gross Domestic Product) went from $1.1 Trillion to $14.5 Trillion between 1971 and 2010," says von Greyerz. Since 2010, total U.S. debt has swelled to perhaps $67 Trillion, a number hard to pin down amidst deceptive government bookkeeping.

"So it has taken an increase in borrowings of $50 Trillion to produce an increase in annual GDP of $13 Trillion over a 40-year period," says von Greyerz.

A Paradise of Borrowed Money

"Without this massive increase in debt, the U.S. would probably have had negative growth for the last 29 years," von Greyerz concluded.

Such unrelenting negative economic growth is the definition of being in a prolonged recession or depression.

Our prosperity, according to von Greyerz, has been a credit-funded vacation from reality, an artificial fool's paradise constructed with money and goods borrowed from others. The bills for all this are now starting to come due.

The vast account of good will and trust the world extended to America after World War II has been overdrawn by America relentlessly living beyond its means.

The credit others gave us was overdrawn by politicians who would not stop printing our paper money, which just happened also to be the Reserve Currency on whose stability and reliability the rest of the world economy depended.

Whiskey and Car Keys

We were supposed to be the world's responsible adult superpower. Instead, we gave our self-serving politicians an unlimited credit card to spend. As comic writer P.J. O'Rourke observed, trusting our spendaholic lawmakers this way has been "like giving whiskey and car keys to teen-aged boys."

Uncle Sam in the world's eyes has gone from hero to big borrower to deadbeat as we paid trading partners back with constantly-debased dollars whose value kept falling. This cheated our creditors out of the full value we promised to pay them for their goods such as foreign oil, German automobiles and Japanese television sets.

The United States continues to go deeper in debt by between $1 Trillion and $2 Trillion each year, officially. Unofficially, because we play accounting games to keep the growing liabilities of Social Security and Medicare in some ways off the books – games that would be illegal if done by a private company – our actual long-term debt keeps growing by at least $5 Trillion every year.

Grown-ups should have known better than to live in a world of fantasy money. As Austrian economist Ludwig von Mises wrote: "If it were really possible to substitute credit expansion [cheap money] for the accumulation of capital goods by saving, there would not be any poverty in the world."

The Bottom Line problem is that paper fiat money does not actually make anything real. When people cease to believe its illusion, as sooner or later they always do, its value vanishes.

The Disintegrating Dollar

"I'll soon explain how this shell game has been, and is still being, played," said Patrick.

"A key here is to understand that the Great Depression that began in 1929 never really ended, nor did the Great Recession of recent years."

"And this is because both these great economic devastations are merely symptoms of something much greater, longer and more terrible...the frightening forest we live in which most people don't even see because of the trees....the immense catastrophe we have been living with for 100 years."

Patrick looked up from refilling his wine glass to see quizzical looks on Ryan's and Peggy's faces.

"We are living, my dear family, during The Great Debasement, the dying of the money as we have known it," Patrick continued.

"America has been sinking deeper into The Great Debasement for a century, since 1913 when the U.S. Dollar died in the fifth of its nine lives and was replaced by an impostor, a counterfeit currency with no intrinsic value."

"How and why this happened, and what is coming next, are a story you need to hear to help you survive after the dollar is no more."

Living in Debasement

When economic historians use the term "The Great Debasement," they are referring to England in the years 1542-1551. During the reigns of King Henry VIII – who had two of his six wives beheaded – and of his sickly, short-lived son and successor Edward VI, the English Crown within a decade gradually removed more than half the silver from the nation's coins.

In March 1542 fully 75 percent of the average English coin's value was the silver in it. This silver value in coins was reduced to only 50 percent by 1545, then to 33.33 percent by 1546.

By 1551, so much silver had been taken out of English coins and replaced with base metals that their silver content constituted only 25 percent of their face value.

This event so shook the English people that we can still hear its echo today in one of Elizabethan playwright William Shakespeare's most popular masterpieces. "No, they cannot touch me for coining," the Bard's King Lear says. "I am the king himself."

After the 1559 coronation of Henry's second daughter as Queen Elizabeth I, she launched a plan to retire the debased coinage and replace it with more honest money.

"As Fair to the Ignorant"

One of Queen Elizabeth's economic advisors was financier Sir Thomas Gresham. Economists still honor him today for Gresham's Law, the recognition that "bad money drives out good." It holds that when people have a choice of trading in either a solid currency or a debased one, they will stash away the good money for themselves and use the debased money in trade, thereby taking good money out of circulation.

For this reason "good and bad coin cannot circulate together," wrote Gresham in a 1558 letter to Elizabeth. In it he explained to her that her father's and half-brother's "Great Debasements" were the reason that "all your fine gold was convayed out of this your realm."

During the Elizabethan Golden Age, English explorers would settle the North American colonies that became America. Virginia, the birthplace of George Washington and Thomas Jefferson, was named for Elizabeth I, who never married and was called the Virgin Queen.

"Brass shines as fair to the ignorant as gold to the goldsmiths," said Queen Elizabeth I.

Yet where money – the medium we exchange for the precious hours of our lives and the sweat of our brow – is concerned, many people strive not to be ignorant. King Henry VIII had to impose legal tender laws to force sellers and buyers to accept his debased, devalued coins.

Governments, including ours, use this same cynical coercion today with a population that knows precious little about how economies and governments really work. Today's government high schools scarcely teach students home economics – how to balance a checkbook – and certainly do not teach how the politicians and Federal Reserve in our Progressive national government unbalance our money system.

Speaking of debasement, consider the history of the British Pound Sterling. Its name comes from the Anglo-Saxon King Offa of Mercia, who during the years of the late 700s created a unit of money that literally weighed one whole pound of silver. This was the namesake of the British Pound.

With silver nowadays worth around $35 per ounce, today's British Pound is worth a bit less than it was in King Offa's day. Now a fiat paper currency like the U.S. Dollar, the Pound as of October 4, 2012, was worth around $1.61.

The British Pound once ruled the world. In 1992 it could not withstand eccentric financier George Soros, who "broke the Bank of England" and reportedly pocketed $1 Billion by speculating short on the Pound.

Debasement's Secret Keys

England's Great Debasement of the 1500s was a trifle compared to today's new Great Debasement.

Two Tudor kings effectively stole a mere two-thirds of the silver value of the money they required England's subjects to use.

By comparison, during the century since Progressives imposed the income tax and Federal Reserve control over our money, the U.S. Dollar has been so debased that it today has only two pennies of the purchasing power of the 1913 dollar.

This means that 98 percent of the U.S. Dollar's previous value has been taken by the government and the Federal Reserve.

We have all been victims of this greatest expropriation of wealth by sleight of hand trickery in human history. And what do we have to show for it?

Those who save their money, live on fixed incomes, or lend money relying on the government to preserve the dollar's value are the ones hurt most by this systematic, deliberate act of debasement.

Those behind the Great Debasement have taken more than our money. They have undermined our independence by confiscating so much of the fruits of our lives' labors that millions of us can no longer afford to control our own choices or retirement.

They have, as we have written before, broken our legs and then offered us a crutch of dependence on government.

The Stimulus Trap

"So look at the trap Keynesian government planners have put the U.S. in," said Patrick.

"According to British economist John Maynard Keynes, the government or central bank can inject money to stimulate a weak economy."

"And this stimulus could work like magic, Keynes believed, because of what he called a 'multiplier effect.' He and his followers have believed that channeling such money to the poor, who had an urgent need to spend it, increased the velocity of money in the economy. It moved faster from one person to the next, creating a feeling that more money existed, that the economy was improving."

"Because of this imagined loaves-and-fishes effect, Keynes calculated, $1 of stimulus money could generate as much economic growth and hiring as would giving $1.50 to those who were not poor and did not need to spend it immediately."

"I'm confused," said Ryan. "Do the government's economists really believe that they are wizards who can turn $1 into $1.50?"

"Of course they do, Uncle," laughed Patrick. "After all, they conjured that stimulus money out of thin air in the first place....yet you accept it as real. It IS magic."

"These self-appointed wizards believe that humans are easily fooled, and this, sadly, is often true," said Professor Pat. "As *The Inflation Deception* documented, money acts like a drug in both the human brain and in society's body politic. It can impair logical thinking, create financial illusions, and cause irrational behavior."

Dangerous Magic

At least Keynes understood that his stimulus magic was dangerous, that in the hands of a Sorcerer's Apprentice it could easily destroy an economy.

Keynes warned that when the economy revives and prosperity returns, the government must prevent runaway inflation caused by all this magic money conjured from out of thin air – unbacked by any real productivity or goods – by quietly taking this stimulus cash back out of circulation via taxes or other means.

"Keynes also believed that stimulus should be used only when the business cycle was low, not constantly injected as it is now," said Patrick. "Keynes would have told politicians never to raise taxes when the economy was down, because this would only make things worse."

It might be possible to spend our way back to prosperity, but Keynes probably would have called it insane to think that a government can tax our way back to prosperity.

We note that policy makers today need to understand that government action might cause not only a "multiplier effect" but also a "divider effect."

In a 2010 study, economists David and Christina Romer – she for a time was head of President Obama's council of economic advisors – produced evidence that every $1 in added tax during an economic downturn caused a $3 reduction in annual economic growth that lasted for several years.

"Keynesian stimulus was supposed to work, which is why more than $5.4 Trillion of stimulus money was spent to stimulate the U.S. economy from 2008 to 2012," said the Professor.

"Of course we now have almost a century of it being tried, and the evidence is fairly clear that it doesn't really work all that well....at least not in advanced economies where it gets discounted by private trading computers at almost the instant the government spends it."

Keynesian stimulus can work in primitive societies with slow market communications and a limited, static currency supply, according to a recent study by economists from the London School of Economics and the University of Maryland.

"Yet they found that the more advanced an economy becomes, the less 'multiplier effect' government stimulus seems to have," Patrick explained.

"By creating a mountain of paper money stimulus out of thin air, the Fed had made business people afraid that this inevitably would cause a future giant tidal wave of inflation, so they cut back on company hiring and expansion. As a result, this Keynesian stimulus made the economy worse, not better."

"A Flood of Irredeemable Paper Currency"

In 1913, under Democratic President Woodrow Wilson, a new cartel of 12 private central banks called the Federal Reserve, would begin turning America's gold into paper – and soon use that paper fiat currency to reshape and rule the United States through chronic inflation.

The Fed was devised in secret, and secrecy concerning its back room transactions and agreements has always been the Fed's preferred mode of operation.

The Fed's enabling legislation provided that American dollars, which until 1913 were supposedly 100 percent backed by the nation's gold reserves, under the Fed suddenly required only 40 percent reserve gold backing.

Many Republicans in Congress were convinced that the Fed was designed to dethrone the gold standard and empower redistributionist Progressives such as the Silverites who wanted easy money, cheap credit that could be repaid with devalued dollars, and government spending unchecked by gold.

One lawmaker who opposed the Fed's creation was conservative Republican and longtime Massachusetts Senator Henry Cabot Lodge – whose son also became a Senator and, in a losing 1960 race, Vice President Richard

Nixon's vice-presidential running mate.

"The [Federal Reserve Act] as it stands seems to me to open the way to a vast inflation of the currency," Lodge prophetically warned in 1913.

"I do not like to think that any law can be passed," said Lodge, "that will make it possible to submerge the gold standard in a flood of irredeemable paper currency."

"Inflation and Deflation Work Equally Well"

Another fierce opponent was Swedish-born Minnesota Republican Congressman Charles A. Lindbergh, father of the later-to-be-famous aviator.

"This [Federal Reserve] Act establishes the most gigantic trust on Earth," said Rep. Lindbergh.

"When the President signs this bill, the invisible government by the Monetary Power will be legalized," he said. "The people may not know it immediately but the day of reckoning is only a few years removed.... The worst legislative crime of the ages is perpetrated by this banking bill."

In 1913 Lindbergh authored *Banking and Currency and the Money Trust*.

In 1917 he published *Why Is Your Country At War and What Happens to You After the War and Related Subjects*, in which he argued that international financial interests had dragged America into World War I for their own benefit. In 1918 government agents destroyed its printing plates.

"The financial system....has been turned over to the Federal Reserve Board," wrote Lindbergh. "That board administers the finance system by authority of....a purely profiteering group."

"To cause high prices, all the Federal Reserve Board will do will be to lower the rediscount rate..., producing an expansion of credit and a rising stock market," wrote Lindbergh. "Then when...business men are adjusted to these conditions, it can check...prosperity in mid career by arbitrarily raising the rate of interest."

The Federal Reserve, warned Rep. Lindbergh, "can cause the pendulum of a rising and falling market to swing gently back and forth by slight changes in the discount rate, or cause violent fluctuations by a greater rate variation and in either case it will possess inside information as to financial conditions and advance knowledge of the coming change, either up or down."

"This," he continued, "is the strangest, most dangerous advantage ever placed in the hands of a special privilege class by any Government that ever existed."

"The system is private, conducted for the sole purpose of obtaining the greatest possible profits from the use of other people's money," said Lindbergh. "They know in advance when to create panics to their advantage. They also know when to stop panic."

"Inflation and deflation work equally well for them," Lindbergh warned, "when they control finance."

When he died in 1924, Lindbergh had left the Republican Party and joined the Minnesota Farmer-Labor Party to run for Governor as its standard-bearer.

"Moneyed Vultures"

Voices reminiscent of President Andrew Jackson's continued to challenge America's new national bank. In a 1932 speech, the longtime chairman of the House Banking and Currency Committee, Representative Louis McFadden (R-Pennsylvania) looked back on almost two decades of America's experience having our currency controlled by the Federal Reserve:

"[W]e have in this Country one of the most corrupt institutions the world has ever known. I refer to the Federal Reserve Board and the Federal Reserve Banks, hereinafter called the Fed."

"The Fed has cheated the Government of these United States and the people of the United States out of enough money to pay the Nation's debt," said Congressman McFadden. "The depredations and iniquities of the Fed has cost enough money to pay the National debt several times over."

"This evil institution has impoverished and ruined the people of these United States, has bankrupted itself," said McFadden, "and has practically bankrupted our Government. It has done this through the defects of the law under which it operates, through the maladministration of that law by the Fed and through the corrupt practices of the moneyed vultures who control it."

"Some people think that the Federal Reserve Banks are United States Government institutions. They are private monopolies which prey upon the people of these United States for the benefit of themselves and their foreign customers; foreign and domestic speculators and swindlers; and rich and predatory money lenders."

"In that dark crew of financial pirates there are those who would cut a man's throat to get a dollar out of his pocket," Congressman McFadden continued; "there are those who send money into states to buy votes to control our legislatures; there are those who maintain International propaganda for the purpose of deceiving us into the granting of new concessions which will permit them to cover up their past misdeeds and set again in motion their gigantic train of crime."

McFadden's views are harsh. Yet they are the views of a professional banker, past Treasurer and President of the Pennsylvania Bankers' Association, and, as we noted above, a Member of Congress who from 1920 until 1931 chaired the House Banking and Currency Committee.

Politicizing Money

The rationalization for creating the Federal Reserve was that it got politics out of decisions about America's money supply. Such decisions would henceforth be made objectively by non-politicians uninvolved in partisan buying and selling of votes. The Fed's bankers would be free to do what is best for the country, not the party in power at the moment.

The Fed was supposed to give us separation of money and state because, despite its name, the Federal Reserve is scarcely more a part of the Federal Government than is the private shipping company Federal Express.

"The Federal Reserve Banks are not federal instrumentalities..." said the ruling in Lewis v. United States 9th Circuit Court in 1992.

The United States Budget in 1991 and 1992 affirmed in passing: "The Federal Reserve banks, while not part of the government...."

The trouble was, and is, that the Fed's seven-member Board of Governors is regarded as a Federal Government agency. Its members are appointed by the President of the United States and consented to by the U.S. Senate for staggered 14-year terms.

Out of these seven, the President appoints the Fed Chair, currently Ben Bernanke, and Vice Chair, currently Janet Yellen, and can reappoint them to a succession of four-year terms. Mr. Bernanke's current term as Chairman ends in January 2014 and as a member of the Board of Governors in January 2020.

Fed Fiddling

"It is wholly impossible for a central bank subject to political control, or even exposed to serious political pressure, to regulate the quantity of money in a way conducive to a smoothly functioning market order," wrote Nobel-prizewinning economist Friedrich A. Hayek in his classic book *Denationalistion of Money*.

"A good money, like good law, must operate without regard to the effect that decisions of the issuer will have on known groups or individuals," wrote Hayek.

"A benevolent dictator might conceivably disregard these effects," Hayek concluded. "No democratic government dependent on a number of special interests can possibly do so."

The Fed originally had one mandate – to protect the value and stability of the nation's money. In 1979 it was given a second mandate – to carry out monetary policy that produces full employment.

Levitating the Market

Some analysts now say the Fed has taken upon itself a third mandate – to boost, or in the words of one financial pundit, to "levitate" – stock market prices.

A few commentators hint that Mr. Bernanke's Fed may have a fourth unspoken mandate – to make the economy seem as good as possible in the months leading up to November 2012, thereby indirectly – and of course unintentionally – to help re-elect President Barack Obama.

Would nominally-Republican Chairman Bernanke have any reason to politically tilt the economy? The 2012 Republican presidential standard-bearer, former Massachusetts Governor Mitt Romney, told reporters that he would not reappoint Bernanke as Fed Chairman in 2014.

While 2012 election-year politics stymied political agreements among the majority-Republican House of Representatives and Democrat-controlled Senate and White House, the Federal Reserve has involved itself in so many government activities that critics charged it has become more like a "Central Planner" than America's politically-neutral Central Bank.

"The Fed has crossed a bright line," warned University of Chicago Booth School of Business Professor John H. Cochrane in August 2012. By imposing stress tests on banks, lending directly to non-banks such as insurance giant AIG, and acting as a regulator, the Fed has greatly expanded its power beyond monetary policy.

The Fed was supposed to have independence from undue political influence as a central bank that dealt only in monetary policy. As Fed tentacles intrude into realms of congressional political power, including regulatory and fiscal policy matters, the Fed cannot expect elected politicians to let it usurp their areas of authority without a response, Cochrane argues.

From its point of view, the Fed might see itself much as the federal courts did vis-a-vis civil rights. When political gridlock and political fear thwart congressional response to important issues, the Fed, like the courts, may feel tempted to fill the policy vacuum created when politicians do nothing.

To complicate this further, when Democrats controlled both houses of Congress prior to the 2010 midterm elections, they created a new agency, the Consumer Financial Protection Bureau (CFPB) to regulate lenders. The funding for CFPB was by law to come directly from the Federal Reserve.

The real power of Congress is in its control of government purse strings. By circumventing congressional power to cut funds for this new agency, this set a precedent for establishing a whole new level of government that cannot be held accountable by the people's elected lawmakers. This is one of many recent expansions of power for the Federal Reserve.

By Their Fruits

The Bible says that we should judge a tree by its fruits. What financial fruits has nearly 100 years of Fed meddling with America's money supply produced?

You can see where the Fed has taken us with a glance at the Federal Reserve Notes most people now call money. When the Fed began, a $50 gold certificate included the words "Will Pay to the Bearer on Demand $50," making it clear that this piece of paper was merely a promissory note held in lieu of real money, genuine money being a fixed and convertible quantity of gold.

Before the Fed became overseer of America's money, the economy rose and fell, with dollars slightly gaining or losing purchasing power. Overall the dollar grew in strength and value, so that what cost $100 in 1829 could be purchased for only about $64 in 1913.

The gold dollar was not only a reliable store of value, but also an excellent investment in an appreciating asset.

Prior to the Fed, stock market panics (often triggered by speculators and corner-the-market schemes) and runs on banks (especially before bank accounts were insured) happened. The economy was resilient, however, and markets usually bounced back quickly to reach new highs.

Down Goes the Dollar

After the Fed began to tighten its stranglehold over our money, the path of the dollar has been almost entirely downhill through inflation. Today's inflated dollar has the purchasing power of only two 1913 pennies, a scant 2/100ths of the 1913 dollar.

During World War I President Woodrow Wilson effectively took the dollar off the gold standard by making it extremely difficult to convert dollars into gold.

After World War I the United States quietly transferred nearly $1 Billion in gold to Great Britain as a gesture of personal friendship between central bankers, a gesture American consumers paid for in lost dollar value, in inflation.

During the 1920s, concluded economic historian Murray Rothbard, the Federal Reserve did not tighten money and bring on the Great Depression as used to be believed. On the contrary, the Fed encouraged inflation to help devalue the dollar in order to make Great Britain's Pound more valuable and ease that nation's return to a gold-exchange standard.

The personal friendship between a top Fed executive and a British policymaker that produced this Britain-boosting tilt in U.S. monetary affairs is discussed in Liaquat Ahamed's 2009 bestseller *Lords of Finance: The Bankers Who Broke the World.*

The high price Americans paid for this Fed debasing of their currency for Britain's benefit provided little of lasting help to us. In 1931 the British Government abandoned the gold standard and has not returned to it.

Conservative Republican Presidents Warren G. Harding and Calvin Coolidge brought America back quickly from a sudden sharp recession in 1920-21 by slashing government spending and the size of the Federal Government.

These conservatives were succeeded by Progressive Republican Herbert Hoover, who as head of a global relief organization funneled food aid to Boleshevik regions of the Soviet Union, which helped prevent the collapse of its Marxist dictatorship. Hoover was invited by Woodrow Wilson to be the Democratic nominee for President, but refused. Hoover declared that he did not favor *laissez-faire* (free market) capitalism, and instead advocated an economy run by voluntary partnerships between government and corporations.

Under Herbert Hoover, the same Federal Reserve System that had expanded America's paper money supply during the 1920s suddenly contracted the money supply during 1929's fever of Fed-stimulated margin stock buying. Unwise Fed manipulation of the money supply, Rothbard argued, is what tanked the Stock Market and plunged America into the Great Depression in 1929.

The Fed's and Hoover's ill-advised attempts at economic engineering, as Milton Friedman later documented, turned what probably would have been just another short recession into a Great Depression that government interference in the economy made worse and prolonged for more than a decade.

FDR's Gold Grab

Among Progressive President Franklin D. Roosevelt's first acts as President was to issue Executive Orders making it illegal for ordinary Americans to own gold bullion, and confiscating people's bullion gold coins (but not coin collectors' numismatic gold coins). Other 20th Century rulers reportedly would outlaw gold ownership – among them Hitler, Mussolini, Stalin and Mao.

In the U.S., bullion was forcibly exchanged for Federal Reserve currency at just over $20 per Troy ounce of gold.

Immediately after this expropriation, FDR arbitrarily raised the official exchange rate of gold to $32 per Troy ounce, then ultimately to $35 per ounce, with government immediately pocketing the value difference to fund his welfare programs and public works projects.

By Executive Order, FDR also outlawed what had become traditional in business contracts – a "gold clause" specifying that the amount to be paid was either in U.S. Dollars or a particular quantity of gold, whichever at the time of payment had greater market value.

These gold clauses in effect secured payment of contracts in either of two different kinds of "money," one of which the government before FDR could not control. (The U.S. Government had banned private transactions denominated in foreign currencies in 1857.)

While nations with freer economic systems rapidly recovered from this global economic downturn, the United States under Roosevelt's collectivist policies wallowed in high unemployment and a stagnant economy.

Pearl Harbor and World War II empowered FDR to conscript unemployed men into the military. Approximately two out of every three American soldiers in this war were drafted, and more doubtless joined the Navy, Marines or Coast Guard to avoid conscription into the Army.

The wartime economy regimented us into a command economy, complete with rationing and austerity on the homefront. Rosie the Riveter and many thousands of other women became skilled factory workers. America, a manufacturing superpower, went back to work.

America still had a work ethic. War production healed us and restored our optimism and self-confidence. Recovery was expected and therefore happened.

By war's end, America's ratio of debt to Gross Domestic Product was at least 122 percent, by some measures even deeper in debt than we are today at close to 100 percent.

Sole Superpower

The United States, however, was still standing, with its factories and cities intact while the world's other once-powerful nations had been knocked to their knees and were severely damaged, weakened and impoverished.

The steps that President Harry Truman took at this history-turning moment, as America became a nuclear superpower, continue to shape our world today.

A 1944 treaty called the Bretton Woods agreement provided that the United States Dollar would continue to be pegged at $35 per Troy ounce of gold, a convertibility redeemable mostly by European central banks. Other major nations in turn agreed to peg their currencies to the dollar, thereby creating what was supposed to be at least the faint shadow of the pre-World War I gold standard.

What we called a gold standard, however, was actually the European prewar "gold-exchange standard." To turn dollars into gold, central banks had to bundle thousands of dollars and in exchange receive gold bars too heavy to carry in a purse or pocket.

This new "gold exchange standard" was impossible for most ordinary people to use. One could no longer take twenty $1 bills to the local bank and swap them for a $20 gold piece as under the pre-Fed gold standard. This did not matter to average Americans anyway, because the prohibition against their owning gold remained law until the mid-1970s, after the effective end of Bretton Woods.

Out of Bretton Woods came two enduring institutions: what today we call the World Bank, by custom run by an American, and the International Monetary Fund or IMF, supposedly the world's "lender of last resort," by custom headed by a European.

The Secret Tax

The United States has done a slow-motion version of using inflation as a de facto tax for a long time.

In 1945 Beardsley Ruml, the Chairman of the Federal Reserve Bank of New York, reportedly delivered a remarkable speech before the American Bar Association in which he declared: "The necessity for a government to tax in order to maintain both its independence and its solvency is...not true for a national government."

"Two changes of the greatest consequence have occurred in the last twenty-five years which have substantially altered the position of the national state with respect to the financing of its current requirements," said Ruml.

"The first of these changes is the gaining of vast new experience in the management of central banks," he said.

"The second change is the elimination, for domestic purposes, of the convertibility of the currency into gold."

FDR's prohibition of gold ownership by Americans meant that U.S. citizens could no longer seek a safe haven here against inflation by converting their paper dollars into gold bullion.

Ruml's speech, published in the January 1946 issue of the quarterly journal *American Affairs*, offered a surprising new vision of what taxes "are really for."

The primary purpose of federal taxation listed by Ruml is "As an instrument of fiscal policy to help stabilize the purchasing power of the dollar."

The implication of Ruml's speech is that government can fund itself merely by printing as much money as needed. Indeed, the *American Affairs* article was titled "Taxes for Revenue Are Obsolete."

The newly-printed dollars produced out of thin air acquire their value, in effect, by devaluing the old dollars that people have earned and saved through their productive efforts.

To prevent a blaze of high inflation or hyperinflation from burning up the entire value of every dollar in circulation, new and old, taxes are used to selectively claw back money from targeted individuals, groups and industries.

Tax policy, said Ruml, will "express public policy in the distribution of wealth and of income, as in the case of the progressive income and estate taxes....[and] in penalizing various industries and economic groups...[and] to isolate and assess directly the costs of certain national benefits, such as highways and social security."

Ruml Touches You Today

Ruml, incidentally, touches working Americans' lives every payday. This Progressive advisor to Presidents Herbert Hoover and FDR devised income tax "withholding" that deducts money from each paycheck before you receive it.

Before Ruml's withholding became Internal Revenue Service policy in 1942, only around seven percent of Americans actually paid income tax – and many who did felt the tax's pain acutely.

Americans were deceptively told that if they signed up to have their taxes withheld, the government would not require most of a year's tax payment from them. In fact, government did collect such tax via withholding without requiring them to write a check.

The government has loved tax withholding from the beginning. It collects money almost immediately from taxpayers. It creates a paper and now-computerized trail of who is paid how much by whom, and establishes witnesses involved on both sides of every income transaction.

Best of all from the government's point of view, income tax withholding takes money almost invisibly from workers – and leaves many euphorically grateful to Uncle Sam, "intaxicated" by their tax refund.

All that such tax refunds mean, of course, is that people overpaid to avoid tax penalties, and gave government interest-free use of their money for a year.

As President Ronald Reagan and other withholding critics have argued, citizens should be required to write one huge check to the IRS every year so that taxes hurt. This, they argue, would give citizens a more realistic sense of their relationship to government than does paying steady, unnoticed withholding taxes out of every paycheck all year long.

For an even better sense of taxation, paying this one huge annual check to the IRS should be required one day before election day, when you vote for the politicians who are taking and spending your money. Tax day should not be as the politicians set it up, with April 15 as far away as possible from November election days in the calendar.

Inflation as Taxation

Inflation also boosts other taxes linked to prices. A 10 percent tax on a $5 product nets the government 50 cents. After inflation doubles that product's price to $10, however, this same tax snares $1 for the government's coffers.

Under a Progressive tax system, few taxes discount the value that taxpayers lose to the government's deliberate inflation. Thus, salary increases that merely offset inflation can push a taxpayer into a higher marginal tax bracket.

Several of the new taxes in the new health law popularly known as Obamacare were deliberately not indexed to offset inflation.

The government thus taxes citizens through the inflation that government deliberately creates. Government then collects even more Progressive taxes when employers pay you more to offset this rising cost of living, because your "rising" income raises the amount of tax you owe. This may push you into a higher tax bracket – even though your real, inflation-adjusted income has not really increased at all.

The Federal Government has thus devised multiple ways to profit from the continuing Great Debasement of the U.S. Dollar.

Lucy and Charlie Brown

Remember that, as we have written before, the income tax was sold to the American people as applying mostly to "rich people." This is the same selling technique of envy that Progressive politicians used to sell Americans the Alternative Minimum Tax (AMT) – that it would hit "only millionaires and billionaires" who were avoiding their "fair share" of taxes.

Thanks to the Great Debasement of the dollar and the inflation it has produced, roughly half of working-age Americans now pay income tax.

The tax increases scheduled for January 2013 target at least 37 million Americans to pay the Alternative Minimum Tax. Yet all of North America – Mexico, Canada and the U.S. combined – has as few as 3.3 million "millionaires" (with net worth of $1 Million, not counting the value of their primary residence), according to the "World Wealth Report" commissioned by the Royal Bank of Canada's RBC Wealth Management, as well as other sources.

As few as 41,200 North American individuals have net worth of more than $30 Million. According to *Forbes* Magazine, as of 2012 the United States has only 425 billionaires.

And now, like Lucy teeing up the football for naively-trusting Charlie Brown to kick, Americans are being asked to vote for Progressive politicians who will impose much higher taxes on the wealthy. How much longer will people fall for this political con game?

At what point will Americans wake up and recognize that our government has a policy of deliberate inflation that sooner or later will have all of us bringing in the number of dollars that only the wealthy used to earn.

Those dollars you earn – so many more than your parents and grandparents did – will be worth only a tiny fraction of what the dollar could buy in 1913 or 1953 or 1983, yet this will not matter. You will be required to pay the tax rate of a millionaire, just as a large fraction of middle-income Americans are paying today in income taxes.

Yet as we are about to discover, the Federal Government does not tax us to get revenue. This is unnecessary. The government taxes us for very different reasons.

"Government is the only institution
that can take a valuable commodity like paper,
and make it worthless by applying ink."

– Ludwig von Mises
Austrian economist

"...Today you and I are working for intrinsically worthless
paper that can be created by bureaucrats — created
without sweat, without creative ability, without work,
without anything but a decision by the Federal Reserve.
This is the disease at the base of today's monetary system.
And like a cancer, it will spread until the system
ultimately falls apart. This is the tragedy of the great lie.
The great lie is that fiat paper represents a store of value,
money of lasting wealth."

– Richard Russell
Dow Theory Letters

"It is well enough that people of the nation
do not understand our banking and monetary system,
for if they did, I believe there would be a revolution
before tomorrow morning."

– Henry Ford

Chapter Five
The Age of
"Modern Money"

"Anyone can create money.
The problem lies in getting it accepted."

– Hyman P. Minsky
Economist

How do you take away a nation's dollars that are as good as gold, and then force its citizens to accept instead paper fiat dollars worth no more than self-serving politician promises?

"New York Fed President Beardsley Ruml's 1945 revelation gives away much of the grand game the money masters and Progressives have been playing for the past 100 years," said Professor Pat.

"The plot thickens," said Ryan.

"Or sickens," said Peggy.

"I told you that the Federal Government had at least three reasons for taxing us, yet that raising revenue was not one of them," said Patrick. Ryan and Peggy nodded.

"Now pay careful attention as I move these three cups, and see if you can tell me which one the pea is under, as I show you how one of the tricks, the illusions of the Great Debasement of modern money, is and has been done."

Imposing Debasement

The opening words of the 1913 law that established the Federal Reserve System specified that one of the Fed's chief purposes was to "furnish an elastic currency," a supply of money that could expand or contract, could stretch or shrink, as the economy – or as those people most closely allied to the ruling politicians – wished.

Most American politicians have hated the gold standard because it makes money solid and "inelastic." A gold-backed dollar's value cannot easily be manipulated for politics or for speculators or special interest group profit.

Gold gave the ordinary people far too much power over the government and the economy, and Progressive politicians in both major parties were determined to grab that power for themselves.

The Federal Reserve's declared purpose was to protect the value of the U.S. Dollar. Under the existing gold standard, however, the dollar had the intrinsic value of gold built into it.

Economists describe such money as having "no counter-party risk," no need for anything beyond itself to create or secure its value. Fiat money, by contrast, depends on government authority to make people accept it.

The gold-backed dollar could protect itself because if people sensed that the government was printing too many, they could immediately demand and receive gold for their dollars. No money-manipulating Federal Reserve System was needed so long as the gold-standard dollar was as good as gold.

The apparent real mission of the Fed was to phase out America's gold-backed dollar as rapidly as the public and the marketplace would allow, and to replace it with a new, modern "elastic" fiat currency backed only by "the full faith and credit of the United States."

"This new money can be multiplied or made scarce whenever those running the Fed wish," said Patrick. "They can put a magic pea under every cup in the game or, *abracadabra*, make it vanish in all of them."

Modern Monetary Theory

How and why the Progressives accomplished this sleight of hand replacement of gold dollars with fiat modern money, beginning in 1913 with their creation of the Federal Reserve System and income tax, can be explained by Modern Monetary Theory (MMT).

This is an analysis of today's economics that Hyman Minsky influenced, and that today's politicians would like you never to learn about – because in key ways it exposes their game.

Most of Modern Monetary Theory's ideas are not new, as its advocates acknowledge.

What is new is how scholars such as University of Texas-Austin economist James K. Galbraith (son of renowned Canadian-born Harvard economist John Kenneth Galbraith, who was FDR's price-controls czar during the Great Depression and John F. Kennedy's Ambassador to India) and University of Missouri-Kansas City economist L. Randall Wray, the author of *Modern Money Theory* and *Understanding Modern Money*, have combined these ideas to reveal a shocking new picture of how fiat modern money economies work.

The cornerstone idea of Modern Monetary Theory is that money is a "creature of the state." All modern fiat money, including the U.S. Dollar, is created by the central government to serve its purposes.

The money supply remains the government's de facto property, MMT implies, just as all the land of a country continues to belong to the government, which will confiscate it if users fail to pay "rent" to the government in the form of property taxes.

The Federal Government likewise routinely exerts its control over the U.S. supply of "elastic" money, expanding it by spreading stimulus dollars or shrinking its supply via taxation that reduces the amount of money in private hands. By such methods, government and its "Creature from Jekyll Island," the Federal Reserve, can modulate money's purchasing power up or down.

Author G. Edward Griffin gave the Fed this nickname after the place where it was planned at a secret meeting of the banking world's and government's top moneymen – Jekyll Island in Georgia.

Three Secrets of Taxation

The U.S. Government has no need to tax anybody to get money, according to MMT. Government can simply produce as much as it wishes, whenever it wishes, either in paper at a cost of two pennies per note, or by having the Federal Reserve add a few zeroes to its computerized accounting.

And yet the Federal Government *does* tax us, and the most productive of us quite heavily, for at least three major reasons:

The first reason government taxes us heavily: Taxation is power.

As Chief Justice John Marshall said in his 1819 ruling in *McCulloch v. Maryland*, "[T]he power to tax involves the power to destroy...the power to destroy may defeat and render useless the power to create...."

Unequal "Progressive" taxation, selectively enforced, can be used to weaken free market foes and reward the government's ruling party allies.

In former New York Fed President Beardsley Ruml's words, unequal taxation can be used to "express public policy in the distribution of wealth and of income," that is, to redistribute a society's wealth as its rulers wish.

Under the pretext of needing revenue, a Progressive-run Federal Government can expropriate money from some people and groups, and transfer that money to others.

Wealth no longer flows to, or stays with, those who succeed through voluntary transactions in the free marketplace. In a Progressive society, whatever portion of it the government desires is seized from its producers by the threat of force by government, which then redistributes this money based on political and ideological considerations.

"19 for Me"

"There's one for you, 19 for me," says the Beatles song "Taxman," inspired by the British government's tax confiscation of exactly that much, 95 percent, of their early earnings. "Be thankful I don't take it all," their taxman sings.

By one estimate, an average baby born in the United States today will either live out his or her life exempt from most direct taxes on the dole, or pay an average total marginal tax rate of around 83 percent in combined direct and hidden taxes, as well as government mandates and the invisible tax of deliberate inflation, imposed even on those who are not as successful as the Beatles.

If Progressives remain in political power, this combined lifetime tax burden will almost certainly exceed 83 percent.

Many of you reading this right now already unwittingly are paying this much. Most people simply have no idea how crushing and deadening a price they pay to the grasping "invisible hand" of greedy Progressive government.

Worse, in Social Darwinist terms, "Progressive" taxation is a selective pressure that reduces the economic survival chances of certain people and increases the survival prospects of others.

Taxation as Manipulation

The second reason government taxes us heavily: Taxation can be used to re-balance an economy afloat in excess liquidity.

Some things, however, cannot be changed by even the most powerful government politicians. They cannot, for example, repeal the law of gravity. Millions of us will have to keep dieting and counting our calories.

And those ruling our government cannot repeal the law of supply and demand. If they double the number of dollars in the economy by dumping trillions of fiat paper dollars created out of thin air, they must soon either take back most of this excess money or risk price inflation which will make everything cost twice as many dollars as before.

This apparently is what Ruml meant when he said that taxes were "an instrument of fiscal policy to help stabilize the purchasing power of the dollar."

The Third Reason government taxes us heavily: Taxation is used, above all, to legitimize and compel citizens to accept a nation's paper fiat currency.

This explains why Progressives rammed through both the new Federal Reserve money system *and* the new progressive income tax in the same year, 1913. This was not a coincidence.

The new income tax and new Federal Reserve money, according to Modern Monetary Theory, would work together synergistically.

Kings have for thousands of years required their subjects to acquire and use the state's usually-debased royal coins or currency for specific purposes.

The Insight of Adam

The founding father of modern economics, Adam Smith, wrote of this in his 1776 masterwork *An Inquiry into the Nature and Causes of the Wealth of Nations:*

"A prince, who should enact that a certain proportion of his taxes should be paid in a paper money of a certain kind, might thereby give a certain value to this paper money, even though the term of its final discharge and redemption should depend altogether on the will of the prince."

A government, in other words, can give value to a new fiat paper currency merely by requiring that people pay their taxes and other government obligations, such as fines and fees and tariffs, only in that currency.

If the government refuses to accept any other form of payment for taxes, then citizens have no alternative except to do whatever they must – work, beg, borrow or steal – to obtain the needed quantity of that new official currency to comply with the tax laws.

What about that nearly-half of the working-age population of the United States that pays no income tax? They, too, have incentive to use the government's paper fiat dollars. This is the currency with which government pays welfare and other benefits.

Those from whom non-taxpayers buy things need the official currency to pay *their* taxes, so they are willing to sell goods and services in exchange for U.S. Dollars.

Money as Coupon d'Etat

Modern Monetary Theory grew out of an earlier school of economics called Chartalism, whose name derives from the Latin word *charta*, meaning a "token or ticket," because it sees paper fiat money first and foremost as a government-issued ticket or coupon required to pay debts to the government.

Modern Money, according to MMT, is at its core a social debt relationship between the government and its subjects. This is the real world basis of Modern Money's value and the control over people that government exercises through it.

The use of U.S. Dollars between private individuals means little to the government, according to Modern Monetary Theory, except insofar as the acts of private buying and selling can be taxed and regulated.

Yet here, too, the value of government fiat money in non-government transactions exists, says MMT, primarily because government transactions require dollars to pay government obligations. This is what gives private users reason to believe that others, who like themselves must pay taxes and other government costs, will accept dollars at a particular exchange value.

How can government increase the value of its paper dollars? Raise taxes.

Higher taxes increase the number of dollars that tax-paying citizens must obtain each year.

The Federal Government can thus increase the supply of its fiat money whenever it wishes merely by printing more, according to MMT.

The government can then offset the risk of inflation by balancing this (and increasing social productivity) by raising taxes in one way or another.

This increases market demand for its money, and it absorbs inflation-causing excess liquidity back into the government whence all money came.

Paper fiat money has long jokingly been called "Monopoly Money," after the cheap paper bills used in the board game Monopoly. The reality, as Modern Monetary Theory reveals, is that today's fiat dollars are literally monopoly money whose value comes from government requiring their use to pay government obligations while maintaining a legal government monopoly on the making of U.S. Dollars. Our dollars really *are* monopoly money, and this grim joke is on us.

Pumping People

This money-tax pump lets the government cheaply extract for its benefit vast amounts of labor value from its subjects.

Our government taxes us via taxation, and also via the hidden tax of inflation as the purchasing power of the money we work so hard to earn and save keeps shrinking, at least for us.

In *The Inflation Deception*, we explained how inflation has become America's Number One export. Foreign central banks buy trillions in U.S. bonds and Treasury notes.

Two out of every three paper U.S. Dollars are held outside the United States, in many cases by people stuffing them into their mattresses in case their own even-less-reliable national fiat currencies break down.

Taxing the World

By deliberately over-printing and inflating the dollar, our government in effect sucks the value out of existing dollars not only inside the United States but also beyond our borders.

This gives us a way to tax the entire world. When other nations' citizens invest in acquiring our dollars, they take on the invisible tax of our government's deliberately-nurtured inflation.

Those who direct America's monetary policies seldom discuss this. They also do not describe themselves as acolytes of Modern Monetary Theory, which casts far too harsh a light of reality on what they are doing. Most nowadays describe themselves, if at all, with the label: Keynesians.

"We're All Keynesians Now"

In 1883, the same year Karl Marx died, in England an upper-middle-class Cambridge University economics teacher and his wife had their first child, a son they named John Maynard Keynes. He would grow up to be the most influential economist of the 20th Century.

In that century when the world was dividing between free market capitalists on one side and collectivists – Marxists, Progressives, socialists, Fascists and Nazis – on the other, Keynes became the symbol of a third way between those two poles.

"We're all Keynesians now," Dr. Milton Friedman, later to win the Nobel Prize in economics, was slightly misquoted as saying in a 1965 *Time* Magazine cover story about the intellectually-fashionable Keynes. President Richard Nixon would later describe himself as a Keynesian.

Riding the Business Cycle

The key to prosperity, Keynes concluded, was keeping an economy's "aggregate demand" high. It scarcely mattered, Keynes concluded, whether this demand came from the private sector or government.

When such demand fell during low points on the roller coaster ups and downs of the business cycle, Keynes taught, then government should save jobs and prime the economic pump of the economy by stimulating the economy.

"The right remedy for the trade cycle is not to be found in abolishing booms and keeping us permanently in a semi-slump," wrote Keynes, "but in abolishing slumps and thus keeping us permanently in a quasi-boom."

Such stimulus, said Keynes, could come in the form of tax cuts, like those used to spark strong economic growth by President John F. Kennedy, called by his scribe Arthur Schlesinger, Jr., "America's first Keynesian

President."

Or such stimulus, said Keynes, could come through government spending that injected money directly.

Such spending, ideally, should be in areas that do not directly compete with or take profits from capitalists – such as public works projects, scientific research, and the like.

Misguided Guru

This, from the outset, revealed the naivete of Keynes, who must have understood that such stimulus money is never free....not even when it is conjured out of thin air.

If government taxes to get the money for such stimulus spending, it is taxing capital away from the private sector that could otherwise have been used by private businesses to expand and hire.

If government borrows to do stimulus spending, it is competing with and crowding out private sector businesses that want and need loans. The interest cost incurred from such debt is also ultimately passed on to taxpayers.

If government or its central bank simply prints fiat money to do stimulus spending, it is debasing dollars (or in Keynes' case, British Pounds) earned and saved by private people. This imposes a hidden tax on rich and poor alike, sooner or later, in the form of inflation.

The Divisive Multiplier

As we noted earlier, government stimulus could exert an almost magic leverage on the economy, according to Keynes. When properly targeted, government spending could have a "multiplier effect."

One dollar of money given to a poor person can produce as much stimulus to the economy as $1.50 given to a rich person, Keynes reasoned.

This multiplier effect happens, he believed, because a poor person will immediately spend the dollar, thereby increasing its velocity in the marketplace.

A richer person, by comparison, has no need to spend his money immediately and is therefore more likely to save it.

Saving money slows down its velocity through the economy, and this means that the goods and services it might have bought go unsold, and that the jobs of those who produce these goods and services produce less profit.

The Sin of Saving

"The paradox of thrift" is what Keynes called this. The old traditional bourgeois values had always taught that it was good to be thrifty, to save. Yet if everyone saved, the economy could spiral into recession.

One person's thrift therefore reduces another's income or costs someone else her job because too much saving reduces aggregate demand in the economy, Keynes concluded.

"Remember how I said that Progressives had turned the world upside-down?" asked Professor Pat. "Here's a prime example."

"Saving used to be a virtue. It accumulated the capital for future investment, or as insurance if something bad happened. It taught people to defer their personal gratification, instant pleasure, by doing what made society more prudent, responsible and secure," said Patrick. "Having money in reserve came from society being more reserved...and, ironically, the Fed requires banks to keep at least some currency or gold in reserve for emergencies."

"However, in today's cult of Keynesianism the act of saving has become a sin that harms society and the economy by slowing down the velocity of money moving from one person's hands to another's."

Keynes, like many intellectuals in his generation, had a brief fascination with Marxism after the Russian Revolution. He was also part of the fashionable Bloomsbury Group of writers and artists, some of whom embraced Fabianism and its aim to transform Great Britain into a socialist country not by one huge revolution but by tiny steps, ratcheting up government power one small government program at a time until government domination of the economy was irreversible.

Yet Keynes soon repudiated Marxism. "For better or worse, I am a bourgeois economist," he wrote.

Perhaps Keynes noticed what Nobel laureate free market economist Friedrich Hayek later quoted in his 1944 classic *The Road to Serfdom*. In 1917 the Progressive founder of the Soviet Union Vladimir Ulyanov, known as Lenin, predicted: "The whole of society will have become a single office and a single factory with equality of work and equality of pay."

Two decades later, Lenin's exiled comrade Leon Trotsky wrote: "In a country where the sole employer is the State, opposition means death by slow starvation. The old Principle: who does not work shall not eat, has been replaced by a new one: who does not obey shall not eat."

Unlike most ivory-tower academic economists who understand only abstract theories, Keynes at his death was worth more than $16 Million in today's dollars that he earned through stock and currency speculation. He put his money where his mind was and succeeded as a capitalist.

"Not Thrift But Profit"

"The engine which drives Enterprise is not Thrift but Profit," Keynes wrote.

Whether or not Keynes was a collectivist, his popular economic ideas provided both camouflage and opportunity for collectivists – Progressives, Fabians, Marxists and others eager to expand the size of government and redistribute wealth.

Keynes taught that government stimulus should be used occasionally, during low points in the business cycle, to lift an economy.

Those on the left who now claim the mantle of Keynesianism use their modified version of his theories to justify continuous government stimulus and intervention in what used to be a free-market economy. Not since John F. Kennedy have these Progressives followed Keynes' idea that stimulus is often done best not by Big Government spending but by cutting taxes.

Keynes suggested, as noted earlier, that channeling stimulus money to the poor produces a multiplier effect. Today's collectivists now invoke him to preach that economic health therefore requires raising taxes on businesses and the rich, and transferring the fruits of their labor to the poor.

Keynes, had he not died in 1946, would today be denouncing these collectivists. He knew that the worst thing government can do during an economic downturn is to raise taxes.

Higher taxes, Keynes knew, drive a struggling economy lower and ultimately harm the poor. Those who advocate higher taxes in today's economy are therefore not honest Keynesians.

Keynes also taught that a healthy economy was driven by "animal spirits," by the energy, dreams, hopes and risk-taking of entrepreneurs and investors.

Keynes believed that government stimulus could improve the marketplace by restoring growth when the business cycle was down. After it did this, however, he thought government should step back and let the efficiency and vitality of free enterprise function without government hindrance or undue interference.

Here, too, Keynes was naïve, according to Nobel laureate Friedrich Hayek and other Austrian free-market economists. Even well-intended government interventions in the marketplace with stimulus will always cause economic distortions, produce malinvestments, and shift money to less efficient uses than would have happened without government's heavy hand.

Worse, governments almost never intervene in the private marketplace with pure intentions. Their stimulus money will always make some richer – usually friends and allies of the ruling politicians – and make others poorer.

In practice, what today is called "stimulus" is just a fancy word to cloak ordinary political redistribution of wealth.

All this would be bad enough had Keynes been correct about such stimulus having a multiplier effect.

The dirty little secret of today's "New Keynesianism" – which replaced the Neo-Keynesianism discredited by the unpredicted Stagflation of the

1970s – is that in the real world there is little or no discernible multiplier effect, at least nowhere near the size that Keynes and his cult have claimed over the years, as we discussed in *The Inflation Deception*.

The Cult of Keynes

Keynes apparently was somewhat mistaken about stimulus and its multiplier effect, according to research published in 2010 by economists Ethan Ilzetzki of the London School of Economics and Enrique G. Mendoza and Carlos A. Vegh of the University of Maryland. They studied how it worked in 44 different countries and found that the Keynesian fiscal multiplier can be effective in emerging countries with low debts, fixed exchange rates and closed economies.

However, in advanced nations like the United States with high debt, floating exchange rates and open economies, Keynesian measures are often much less effective, as we noted in Chapter 4.

In today's ongoing Great Recession, Keynesian stimulus *has* worked – but not as expected. Exactly as we predicted in *The Inflation Deception*, it turned into an "anti-stimulus."

Yi Wen, economist and Assistant Vice President of the Federal Reserve Bank of St. Louis, analyzed what effects nearly $2 Trillion of stimulus money injected into the economy by the Federal Reserve has had. What he found surprised him.

Wen's conclusion: this immense stimulus spending frightened business people, who concluded that it would soon generate a tidal wave of destructive inflation in the economy. As a result, the stimulus led them to cut back on hiring and expansion in their companies.

This stimulus did worse than fail – it backfired, producing more harm than benefit and leaving our nation with astronomical debts.

The Stimulus that Failed

The Fed now argues that things would have been far worse had it and the government not injected more than a combined $5 Trillion of stimulus money into the economy, a rationalization akin to President Obama's claim to have "created or saved" millions of jobs. This unprovable gambit allows the powers that be to take credit for every job that somehow survives their policies.

The reality, of course, is that after spending over $5 Trillion to stimulate the economy, economic growth in the Second Quarter of 2012 was an anemic, stall-speed 1.3 percent. Subtract from that real world inflation that is running above seven percent, eating up growth calculated using an understated deflator number, and all this stimulus has given America negative 5 percent growth – a continuing Recession.

If this is all that Keynesian stimulus can do at a cost of more than $5 Trillion, then it is far too expensive a tool for practical use.

In 2009 Mr. Obama's new chief economic advisor Christina Romer predicted the Keynesian "multiplier" for debt-financed stimulus – that every dollar the government borrowed and spent would generate $1.57 in new economic activity.

0.29

In September 2012 the Independent Institute analyzed the Obama stimulus effect from his inauguration on January 20, 2009, through June 30, 2012, the end of the most recent quarter for which good data about the nation's GDP was then available.

During this period the Federal Government borrowed and spent $5.23 Trillion dollars. From December 31, 2008, until June 30, 2012, America's Gross Domestic Product increased by $1.52 Trillion.

What this means, the Independent Institute economists calculate, is that the real Obama multiplier effect fell far short of 1.57 percent. It also fell far short of 1.00, which at least would have been a break-even point, that each dollar spent generated one dollar in economic activity.

The Obama Keynesian multiplier, these experts found, was actually 0.29, only 29 cents of economic growth for every stimulus dollar spent, a "loss" of 71 cents on every federal stimulus dollar.

"Put another way," the Independent Institute reported, "for every $3.43 the

U.S. Government has borrowed [and spent] during President Obama's tenure in office, the nation's GDP has increased by just $1.00 – a lousy rate of return on the government's debt-financed 'investments.'"

This government stimulus money, remember, has an economic cost beyond mere interest payments. It was gotten by taxing businesses, or by "crowding out" the private sector from loan money that private companies might otherwise have been able to get and use for investment and hiring.

The strict new Obama era rules on lenders have made it extremely difficult for small businesses and mortgage-seekers to qualify for loans – and therefore given lenders vastly more incentive to lend to government instead.

A similar politicized tilt against lending to private, and in favor of lending to state-run, companies is now underway in Communist China's "state capitalist" system.

We will never know how many jobs or how much economic growth this $5.23 Trillion might have generated in the hands of the free market – yet it's safe to assume that capitalists would have produced more with it than the government did.

President Obama would later laugh while saying that the "shovel-ready" government jobs he wanted to spend stimulus money on "weren't so shovel-ready." Vast amounts were spent – disproportionately in states and cities that voted Democratic – on crony companies such as Solyndra and on bankrolling state and local governments to preserve public service union jobs.

Mr. Obama increased federal jobs during this period by 11.4 percent, and raised the already-high pay for such government jobs even higher.

While he did this to expand the government sector of society, one in five men aged 25-54 were left without work. Even the deceptively-low official national unemployment level has been stuck for more than 40 months above 8.0 percent. In reality, nearly 25 million Americans are unemployed, underemployed or have quit looking for a job.

As we noted earlier, under President Obama the Federal Government share of America's GDP has risen from 19.5 percent to almost 25 percent – increasing the Federal Government's size relative to the economy by more than 25 percent.

This, it turns out, has been the Progressive's real "multiplier" effect, a 1.25 multiplication of government size, wealth and power in less than four years.

This redistribution of wealth in society under the pretense of a Keynesian "multiplier effect" also has a "divider effect." It pits those who get a government paycheck or welfare check against the rest of us who pay the price for this redistribution.

Magical Mystery Cruise

"We can easily conjure up plausible theories as to what we will do when it comes to our next tack or eventually reversing course," said Richard Fisher, President of the Federal Reserve Bank of Dallas, in a September 19, 2012. speech.

"The truth, however," said Fisher, "is that nobody on the [Fed Open Market Committee], nor on our staffs at the Board of Governors and the 12 Banks, really knows what is holding back the economy."

"Nobody really knows what will work to get the economy back on course. And nobody – in fact, no central bank anywhere on the planet – has the experience of successfully navigating a return home from the place in which we now find ourselves."

"No central bank – not, at least, the Federal Reserve – has ever been on this cruise before," said Fisher. "I believe that with each program we undertake to venture further [into quantitative easing], we are sailing deeper into uncharted waters."

Fisher has been an outspoken critic of the Fed's recent stimulus expansions because, as he noted in this speech, trillions of dollars of private bank reserves and corporate money "are sitting on the sidelines," unspent.

"Why would the Fed provision to shovel billions in additional liquidity into the economy's boiler room when so much is presently lying fallow?" Fisher asked. "One of the most important lessons learned during the economic recovery is that there is a limit to what monetary policy alone can achieve."

Keynesian Quandary

Lord Keynes came from an earlier era in which people were productive and saved – a time long before middle class people lived on credit cards instead of earnings, and long before 35 percent or more of America's Gross Domestic Product (GDP) began coming each year from government spending....a perpetual heavy government stimulus program.

The new analyses of today's spendthrift era of living on credit provide evidence that Mr. Obama's stimulus policies have wasted far more than $5 Trillion on a wrong-headed Progressive version of Keynesian dogma. Obama and Fed Chair Bernanke have created a process that must again and again keep pumping in artificial money in an attempt to stimulate what has now become an artificial U.S. economy.

Many of today's Progressives will never be persuaded by the evidence that they are wrong. For them Keynesianism is a religion that promises them the keys to a heavenly utopia in which government, run by a chosen elite of superior people like themselves, will control all aspects of human society, including the economy.

Other Progressives, says economist Peter Ferrara, do not necessarily believe in Keynesian analysis. Their view is like that of former Colorado leftist Senator Tim Wirth, who once said: "We ought to 'ride' the global warming issue, because even if it proves wrong it will help us to make changes we ought to make anyway," i.e., in the direction of bigger government, higher taxes and more regulation of businesses and individuals.

Some Progressives likewise "ride" Keynesianism, says Ferrara, because this is a train going in the same direction they want to move America – towards bigger, more controlling government, and more redistribution of wealth from rich to poor.

Some Progressives do not care whether massive wealth redistribution is imposed in the name of social justice or Keynesian economic stimulus; whatever gives them the power to do it is all right with them.

As True Believer religious zealots possessed by their own "animal spirits," the actual Keynesian response to all evidence that they failed is: "We didn't spend *enough* on stimulus. If we spend three, five, 10 times more, stimulus will work."

Push *New York Times* Progressive econopundit Paul Krugman's button, and he will automatically repeat this, his sacred chant, a dozen times before pausing for breath.

If only government would tax away everything from the rich and spend another $100 Trillion on stimulus to redistribute that wealth, the Progressives insist, then a glorious new age will dawn.

We have a simple question: even if Keynes were right, and that by fine-tuning government spending and taxing policies we can flatten the business cycle to create what he called a permanent "quasi-boom," would it be wise or in our best long-term interest to do this?

If you could wave a magic wand and transform the world into never-ending springtime, or harvest time, or summer, would you be wise to do so?

Perhaps, as some Austrian economists believe, the business cycle itself is a by-product of government interference producing bubbles in the economy. In a truly free market, some believe, the business cycle, at least in its most extreme swings from boom to bust, might nearly vanish.

Stopping the Economy's Heartbeat

Or maybe what we call the business cycle is a natural part of the economic part of being human. Maybe this is a natural cycle like summer and winter, day and night, or the never-ending circadian rhythm that makes us dream every 90 minutes when asleep or daydream every 90 minutes when awake.

Maybe the business cycle is the heartbeat of human beings interacting with one another in society.

Maybe, like a properly-functioning set of lungs, a healthy economy needs to take in good air by expanding when it inhales, and then contract to let out bad air when it exhales. The economy requires both. If all we did was inhale, the lungs would swell and then burst. We need a full cycle to energize new and effective enterprises, and then to break down and eject enterprises that no longer function efficiently.

If such cycles keep the economy healthy, then using Keynesian stimulus and taxation to flatten the business cycle's highs and lows might weaken the vitality and innovative forces that come from this process of high growth followed by low destruction and renewal – these hot summers and chill winters of the economy.

With some exceptions, great human civilizations have come from temperate climates where the guiding fable from the ancient Greek storyteller Aesop was of the lazy grasshopper and the industrious ants. The ants knew that winter was coming, that sunny days must be used to store up food for an approaching season when crops would stop growing and snow could fall. The ants were capitalists.

Wall Street Casino

"Remember how I told you America is suffering from a risk shortage?" asked Patrick. Ryan nodded.

"A healthy free marketplace must have two things – reward...and risk," said Patrick. "If everything is high risk, most people will likely stop investing. If everything is low risk and high reward, then people will show no prudence or caution in how they invest their capital."

"What happened in 2008?" said Patrick. "Several giant banks apparently had been told by the politicians to whom they contributed lots of money that they were 'Too Big to Fail,' that if things got bad the government would use taxpayer money to bail them out."

"So these banks did the logical thing under such circumstances. They became high rollers," said Patrick.

"When these banks' wagers won, they pocketed their winnings as private profit. When their dice came up snake-eyes or boxcars, however, the Fed and the U.S. Treasury rushed in with trillions in stimulus money to bail them out, to cover their bad bets."

"Progressives are almost always wrong, yet about this they had a point," said Patrick. "The banks were 'privatizing their profits but socializing their losses,' sticking taxpayers with the bill. And most of those giant banks today are 30 percent bigger than they were in 2008....so I guess they're now 'Too Too Big to Fail.'"

"You can call this many things," said Patrick, "but you can't call it free market capitalism. This is crony socialism, welfare for rich gamblers. It's just one more symptom of the Great Debasement."

"It's made a large part of our economy as fat and irresponsible and unethical as the government itself. In fact it largely erases the line between the private sector and government."

"These companies have zero risk, thanks to government, so they have no prudence or sound judgment in how they conduct business," said Patrick. "This is not how those who built America created a strong, resilient, efficient and innovative economy."

"And neither is today's Progressive government that puts its boot on the neck of any company or capitalist who doesn't contribute to the ruling political party, or bow to the demands of petty bureaucrats and Progressive community organizers."

"Today's government has removed market risk for its cronies, but replaced it with the risk and uncertainty of all-pervasive government power, whim and caprice," said Patrick. "This is not what a free country is supposed to be like."

Financial Repression

"Remember when I earlier mentioned 'Financial Repression'?" asked Patrick.

"How could we forget words like that?" said Peggy.

"The banks are part of this, and so is the Federal Reserve," said Patrick. "It's what happens when a central bank such as the Fed pushes interest rates below the rate of inflation."

"If you're a person who prefers to put your money in the bank, where it is safe and earns a little interest, the Fed's current deliberate policy of 'financial repression' means that you are losing money by saving it."

"Inflation, in everyday terms like food and gas prices, now runs around seven percent. This is eating up the purchasing power of your money faster than the one percent or so of interest your bank pays you."

"Why is the Fed doing this? It has several reasons," said Professor Pat.

"First, the biggest borrower in the country is the Federal Government. It now borrows approximately 41 cents of every dollar it spends – 75 percent of it borrowed from the Fed itself. The government, as noted earlier, is borrowing at least $47,000 and as much as $58,000 every second (depending on which ever-changing official numbers we believe) so our spendaholic politicians don't have to cut their spending."

"The Fed is forcing interest rates down to make such borrowing and spending possible. If the government had to pay the historically normal interest rate of around six percent, its borrowing would cost taxpayers at least $6 Trillion more every 10 years," said Patrick.

"By keeping inflation higher than bank interest rates, the Fed can also print money into that gap, the difference between the two – in effect, skimming off for the government the value you should have been paid in interest. It's too complicated to explain in layperson terms right now how this works, but it does," said Professor Pat.

"Moral Hazard"

"What the Fed really wants is to punish savers so badly that people withdraw their savings from the bank and take the 'moral hazard' of putting their money into riskier investments that could boost the economy – things like stocks."

"In fact, the Fed would be delighted if people just stopped saving and spent their money on *anything* to boost consumption, which has been approximately 71 percent of America's recent Gross Domestic Product," said Patrick.

"The Fed apparently would rather have you spend your savings on Chinese-made goods, and send the profit to them, than save it for yourself and slow down the Keynesian economy," Patrick added.

"Remember that Fed Chairman 'Helicopter' Ben Bernanke gained his nickname because he once agreed that his Fed could stimulate the economy by just tossing money out of a helicopter."

"I could use a new laptop computer," offered Ryan.

"We've got a leaky roof to repair," said Peggy, "and we just can't afford both right now."

"You're both what the Progressives call 'economic patriots,'" said Patrick, smiling at his aunt and uncle. "They just want you to spend, even if you have to go deep in debt as your government does."

"I told you that risk is essential for businesses, that it makes them prudent and careful," Patrick continued. "However, ordinary bank savers should not be forced to put their money or their credit at risk, which is what the Fed is doing by practicing financial repression. It's one of the ugly games the Fed is now playing with Americans," said Patrick, "and they ought to stop it....but they won't."

"[F]inancial repression in its many guises...will likely be with us for a long time," wrote economist Carmen M. Reinhart in March 2012. Both the Fed and European central banks now use it not only as a concealed tax but also as a way to impose additional control over banks, pension funds and businesses.

Reinhart is a senior fellow at the Peterson Institute for International Economics and co-author, with Harvard's Kenneth S. Rogoff, of the bestselling 2009 book *This Time Is Different: Eight Centuries of Financial Folly.*

One emerging form of financial repression among European nations, she notes, is that for international bank transfers "the amount of disclosure, red tape and other requirements that are necessary to make such transfers has been on the rise," rationalized as preventing money laundering and tax evasion, yet also amounting "in some cases, to administrative capital controls."

Atlas Shrugs Again

Even if Keynes' theory were correct, a flood of government stimulus paper money is the wrong medicine for what currently ails the American economy. Textbook Keynesian dogma is to use stimulus money to create liquidity to lubricate a dry economy – but today's economy is not dry – it's frozen.

Corporations and banks reportedly have been holding back as much as $3 Trillion on their books, reluctant to hire or invest because of uncertainties about how President Obama's tax, regulatory and other policies will affect their future costs of doing business.

In addition to such uncertainty, some corporate leaders may unconsciously be living out what philosopher Ayn Rand depicted in her novel *Atlas Shrugged* – a strike by society's "productive" people, a work stoppage to protest ever-higher taxes and oppressive mandates that force them to pull the wagon millions of others feel entitled to ride in for free.

We may be seeing Atlas, who in Greek mythology held up the world on his shoulders, shrugging right now in the ongoing slowdown by investors and employers against Progressives who are openly hostile towards business and capitalism.

When this $3 Trillion regains velocity in the economy, and when hundreds of billions of stimulus money now held unspent by recipient entities is put back into play, the economy will be awash in money.

Explosive inflation will likely then send prices soaring, perhaps leading to Post-World War I, Weimar-like hyperinflation in which prices double every month.

By following a Quantitative Easing policy deliberately intended to ignite inflation, as we warned in *The Inflation Deception*, Mr. Bernanke and the Federal Reserve are literally playing with fire in a room full of frozen gasoline.

Keynes on Inflation

Keynes was flawed on some economic matters, such as his approach to stimulus. He also called gold a "barbarous relic," apparently because a gold standard thwarted his approach to economic manipulation by governments and central banks.

Keynes did have a very clear view of some key economic issues, one of which was inflation.

"Lenin is said to have declared that the best way to destroy the capitalist system was to debauch the currency," wrote Keynes in his 1919 book *The Economic Consequences of the Peace*.

"By a continuing process of inflation, governments can confiscate, secretly and unobserved, an important part of the wealth of their citizens...."

"As the inflation proceeds and the real value of the currency fluctuates wildly from month to month," he continued, "all permanent relations between debtors and creditors, which form the ultimate foundation of capitalism, become so utterly disordered as to be almost meaningless; and the process of wealth-getting degenerates into a gamble and a lottery."

"Lenin was certainly right," Keynes concluded. "There is no subtler, no surer means of overturning the existing basis of society than to debauch the currency. The process engages all the hidden forces of economic law on the side of destruction, and does it in a manner which not one man in a million is able to diagnose."

Any resemblance between what Keynes describes and the Great Debasement of American money and values carried out by Progressives is, of course, purely coincidental.

How much has the Great Debasement cost America, at least in money? Using data compiled by the Bureau of Economic Analysis (BEA), *Forbes* columnist Louis Woodhill estimated in June 2012 that, in lost growth and value, the fiat dollar that Progressives imposed on America between the years 1974 and 2010 cost America more than $80 Trillion.

These 36 years, it goes without saying, are only a fraction of the 100 years of the Great Debasement.

If the annual loss over these entire 100 years averaged the same as during the 36 years of Woodhill's calculation, then the Great Debasement in dollars has cost the United States at least $222 Trillion.

Speaking of coincidence, the "fiscal gap" between U.S. projected long-term debt and projected revenues – the debt that could bring America's economy and dollar crashing down in ruins – has been calculated by Boston University economist Laurence Kotlikoff to be $222 Trillion.

Goldilocks Dollars

Progressives found Keynes especially useful in justifying the transformation of America's money from gold-backed dollars into what the enabling legislation of the Federal Reserve called an "elastic currency."

Gold is solid, not elastic, and not the rapidly-evaporating currency economists refer to nowadays when they speak of the Fed conjuring "liquidity" out of thin air.

Gold, therefore, had to be replaced as rapidly as the Progressives could make it vanish. It stood in their way of being able to build an all-devouring government by using unlimited paper fiat money.

Later we shall learn why economist Alan Greenspan, who would later chair the Federal Reserve, said that Progressives had to get rid of the gold standard before they could impose their agenda on America.

Keynes offered a useful key to such transformation.

Traditional money is expected to perform at least three essential functions. It is a unit of account through which people can quantify and agree upon a transaction.

It is a medium of exchange people use to carry out a transaction.

In the past, money was to be a store of value whose worth would not evaporate in the hands, wallets or storage safes of those relying on it.

This third function of old-fashioned money is at odds with Keynesianism, because Keynes regarded saving and thrift as bad for the economy.

The Progressives' logical replacement for gold, therefore, has become what we call Goldilocks Money....money that as Goldilocks, the little girl in the fairy tale, says is "just right."

Goldilocks money – today's U.S. Dollar or the Euro are prime examples – looks like money and exchanges like money and serves as units of exchange like money.

Yet in today's Keynesian world, where the government does not want people slowing down the velocity of money by saving it, Goldilocks money is, by design, increasingly unreliable as a store of value.

This gives people less incentive to save dollars for any length of time. Like Modern Monetary Theory's idea of money as a government coupon, the dollar is becoming like a coupon with a kind of expiration date as its value keeps evaporating. The longer you hold it, the less it is worth.

The greater the risk of inflation eating up the purchasing power of Goldilocks money, the faster people rush to spend, to exchange their paper dollars for something else.

After World War I, when the "Progressive" Weimar Republic in Germany created a brief false prosperity by printing mountains of paper currency, the German people by 1922 began to "smell a rat" in what the government was doing, as we described at length in *Crashing the Dollar*:

If money is a reliable store of value, like gold, human beings tend to hang on to it. If money is unreliable as a store of value, then, as economic historian Jens O. Parsson wrote in his 1974 book *Dying of Money: Lessons of the Great German and American Inflations*, "People naturally wish to hold money less and to spend it faster when they see its value falling."

As Germans began to wake up and smell the socialist Weimar government's hyperinflation of their money, wrote Parsson, "Velocity took an almost right-angle turn upward in the summer of 1922," as people began spending their Marks faster and faster.

In Weimar, as the inflation worsened, in part because of this proto-Keynesian accelerating money velocity, "Nobody wished to retain money," wrote Austrian economist Ludwig von Mises. "Everybody dropped it like a live coal."

According to von Mises, on the German Stock Exchange this behavior was called *Flucht in die Sachwerte*, "flight into investment in goods," the conversion of evaporating paper money into almost anything durable – gold, furniture, antiques – that would retain value better than a politician's promissory note, i.e., paper fiat money like Weimar's Mark or today's U.S. Dollar.

Trapped by Fiat

The Weimar politicians and Reichsbank by summer 1922 were frantically printing more and more paper money, trying to keep ahead of the inflation that their unrestrained printing was causing.

In July that year the government passed a law permitting, under license, local and state governments and even large corporations to issue emergency money tokens to boost the money supply. These Chartalist-Modern Monetary Theory tokens, aptly enough, were actually called *Notgeld*, "not gold."

Today America's Federal Reserve finds itself trapped into issuing what will likely be an endless succession of QE stimulus programs to feed the cravings of the addicts its easy-money policy has created. Weimerica may be closer than they think.

Once inflation and a loss of faith in faith-based fiat money start to take hold in a nation, they can be extremely hard to stop. In his September 2012 speech, Dallas Fed President Richard Fisher noted that disquieting signs of inflation are starting to appear.

This is especially true when a government and its central bank, like ours today, are afraid to stop expanding the money supply, lest the economy go into addiction withdrawal symptoms and crash.

Yes, the Progressives wanted an "elastic" Goldilocks money that people could be forced to accept, yet that was such an unreliable store of value that people would rather spend than save it.

The Progressives got their wish: a dollar as greatly debased as their values.

The issue now, as economist Hyman Minsky raised, is what our Progressive government and Federal Reserve will have to do to make the American people keep accepting it.

Today, as America's economy weakens, the U.S. Dollar becomes evermore vulnerable to reaching a tipping point that could cause other nations to stop using it as the world's Reserve Currency.

Next we consider four potential tipping points – political, natural, military and global – that could lead to a post-dollar world.

"A wise and frugal government,
which shall restrain men from injuring one another,
shall leave them otherwise free
to regulate their own pursuits
of industry and improvement,
and shall not take from the
mouth of labor the bread it has earned.
This is the sum of good government...."

– Thomas Jefferson
First Inaugural Address
March 4, 1801

Part Two
Tipping Points

Chapter Six
The End of the Republic

Tipping Point # 1
Anti-Capitalist Politicians
Take Control of the Government

"When the people find that they
can vote themselves money,
that will herald the end of the republic."

– Benjamin Franklin

"We are becoming a 50-50 nation –
half of us paying the taxes,
the other half receiving the benefits."

– Niall Ferguson
Harvard Economic Historian

The strength and credibility of the U.S. Dollar, which in 1971 ceased to be anchored to gold, now depend on the health of America's economy as well as on the confidence and faith people have in America's fiat paper faith-based currency. On the day this faith collapses, the dollar is toast.

The strength and well-being of America's economy are no longer based on free enterprise, but on government.

Federal Government spending is now roughly 25 percent of our economy's entire Gross Domestic Product. Federal, state and local government spending combined comprise roughly 35 percent of GDP, more than a third of the whole economy. Government regulatory, tax and spending policies shape and control banking and other private sector behavior throughout the economy.

Our government, meanwhile, is driven by politics that are increasingly partisan, ideological and strident.

Many politicians now seem willing to say or do anything to win or cling to power, because our huge, intrusive government has become a grand prize and treasure trove for whomever captures and controls it.

The health and survival of the U.S. Dollar therefore now depend on politics that are increasingly polarized.

We no longer live in a nation with separation of marketplace and state.

Yet mixing politics and the market is, even at best, like trying to mix oil and water. What is good economics and business judgment is usually bad modern politics, which consists mostly in giving away money taken from taxpayers.

Successful politicians nowadays are often those who play this game of polarization and redistribution of wealth most forcefully.

Yet in economist Adam Smith's warning at the beginning of Chapter Three, political manipulation of resources in the marketplace "would nowhere be so dangerous as in the hands of a man who had folly and presumption enough to fancy himself fit to exercise it," who believes that his centralized coercive authority is wiser than the decentralized, voluntary "invisible hand" of the free marketplace.

The politicians who now intrude government into every aspect of business are outraged that business people now feel compelled to invest in opposing their re-election.

When government left business alone, business people had no reason to waste their time or money in partisan politics.

The comic writer P.J. O'Rourke was correct, of course, when he wrote that when government involves itself in buying and selling, the first things bought and sold will be lawmakers.

We now have lots of lawmakers who see their power to threaten to impose higher taxes and heavier regulations as a way to increase their own "market price" for campaign contributions from frightened business people.

What happens to our dollar if politicians who, either by incompetence or ideology, destroy the productivity of America's economy and the value of the dollar? Or if they destroy the American republic itself?

Is this how the U.S. Dollar will die?

Changing Parties

President Barack Obama is the apotheosis, the avatar, the living embodiment of Progressivism.

Mr. Obama's presidency has been a triumphal advance of Progressivism making our government bigger, an advance and entrenchment of its ideology that future elections and politicians might never be able to roll back.

Those who know history find this strange.

As we have seen, it is ironic that Mr. Obama's Democratic Party began as the party of small government, low taxes, and free market individualism. This was the party of Thomas Jefferson and Andrew Jackson, of rural values, slaveholders, and frontier pioneers seeking elbow room away from corrupting big cities.

Back in the 1800s, when an American said he was a "liberal," that meant he favored economic liberty and free markets. This is why some libertarians today describe themselves as "19th Century Liberals."

The once-honorable label Liberal was co-opted and used as a disguise by Progressives eager to abolish free enterprise and replace it with one or another form of collectivism. Today Progressives have debased the word

Liberal just as their Great Debasement has degraded the once-solid U.S. Dollar.

The Republican Party, by contrast, began as a proto-Progressive Third Party of radical reform that sought the abolition of slavery. Its first President, Abraham Lincoln, made the central government in Washington, D.C., supreme over the rights and sovereignty of the states that created the Federal Government. Lincoln created the first, relatively short-lived income tax in America, the first military conscription, and a wartime economy based on fiat Greenback paper currency.

After 620,000 died in the War Between the States, and President Lincoln's assassination, Republicans won the White House and Congress for most of the next 67 years, as the party of industrialization, Wall Street and Progress.

Yellow Dogs

In the wake of this war, the Democratic Party was largely out of national power for two-thirds of a century. It survived in bastions regional or local.

In the South after Reconstruction, embittered citizens taught their children to vote even for a "yellow dog" so long as he was a Democrat. This produced one-party Democratic states for almost a century that became the foundation of Franklin D. Roosevelt's odd-bedfellows New Deal coalition.

In the West, populist farmers and miners became Democrats because they associated Republicans with Wall Street, the gold standard, and railroads that charged high rates.

Big city Democratic machines from Kansas City to Boston recruited recent immigrants from Ireland, Italy and elsewhere to become Democrats, in part by giving them government jobs.

This tradition continues today in the Democratic Party's outreach to Latino immigrants.

Progressives also encourage Hispanic immigration, but their aim is to import something socially-mobile America has always lacked – a Proletariat class that can be propagandized and used as cannon fodder in class warfare. Because people here prior to Progressivism all had the ability to acquire property and improve life for their children, Marxism and its class warfare would not take root in American soil.

Today, however, Progressives are importing their Proletariat from Latin America and destroying American opportunity as fast as they can. This is why you now hear Progressive politicians openly voicing their hatred for capitalism and trying to stir envy and hatred against the successful. They want class warfare and now think they can win it. We shall see if the Left is right in upcoming elections.

Tax and Tax, Spend and Spend

Progressive Democrat Franklin Delano Roosevelt, the only American President ever elected to four terms, based his political success on a simple slogan: "Tax and tax, spend and spend, elect and elect."

The Irish playwright and Fabian socialist George Bernard Shaw likewise proclaimed: "He who robs Peter to pay Paul can always count on Paul's support."

FDR appeared to be grabbing power throughout society. New Deal programs seemed to have money for everybody – Social Security for the elderly, government jobs for the unemployed, cash for farmers, money for writers and artists, and more.

FDR's "National Socialism"

The Democratic Roosevelt, "having promised to give the government back to the states, had seized power at the center," wrote a famed liberal British correspondent in his 1982 bestseller *Alistair Cooke's America.*

"For two dizzy years, America had a fling at National Socialism...," wrote Cooke, who was in the U.S. during these years and in 1941 became a citizen.

"Roosevelt was for all administrative purposes a dictator....," wrote Cooke. "In effect, he signed a check in the government's name for several billion invisible dollars."

After two years of National Socialism American-style, the U.S. Supreme Court somewhat reined in Progressive FDR's dictatorial powers.

By then, millions of Americans were hooked for life on government handouts, a pusher-addict relationship that has endured and grown to this day. Many critics have argued that this addicting flow of free government money and other goodies from politicians amounts to vote-buying with taxpayer money.

Benjamin Franklin warned that this would mean "the end of the republic." It may soon also mean the demise of the U.S. Dollar.

America's Transfer-mation

When John F. Kennedy was President, more than 50 cents of each tax dollar was spent on national defense.

Today nearly two-thirds of every dollar the Federal Government spends goes to "transfer payments," taking money from your pocket and moving it to the pocket of someone else the politicians deem more worthy than you of enjoying the fruits of your labor.

The biggest activity of America's politics today is not protecting the inventors and producers of wealth and prosperity.

Government's biggest activity today in our Transfer-mational, politicized economy is the confiscation of an ever-larger share of what our most productive citizens earn, and the redistribution of their earnings to others more favored by politicians. Many of the recipients produce less with such capital than do those from whom this economic resource is taxed.

A 50-50 Nation

Historians will note that by year 2012, only 51.5 percent of working Americans paid federal income tax. Everyone paid hidden taxes, passed on to them via higher prices, yet the other 49.5 percent who paid no federal income taxes were propagandized to vent their envy and dissatisfaction at "the rich" and the "greedy corporations" whose products they bought.

This privileged 49.5 percent has been encouraged to think of the government as a big goody-dispensing machine that gives them stuff for free.

Kenneth Minogue of the Fabian socialist-founded London School of Economics warns that today's "social democratic" welfare state depends on an ever-narrowing tax base as it seeks to extract more and more taxes from fewer and fewer people and corporations.

Yet once 51 percent of the people in such a society become "public clientage," – welfare recipients – wrote Minogue, the decline of that society inevitably sets in.

America, as *Wall Street Journal* economics writer Stephen Moore observed, has truly become "a nation of takers, not makers."

By 2011, more than twice as many people (22.5 million) worked for local, state or Federal Government than in all of manufacturing combined.

What else would we expect from a nation in which one political party claims to represent the takers and ever-expanding collectivist government, and the other party says it represents the makers, businesses large and small and the self-reliant individualism that built America?

Transforming Parties

In the two decades following World War II, the two major parties generally shared a belief that prosperity was in everybody's best interest. Foreign policy differences ended at the ocean's edge, and the Cold War with the Soviet Union prompted defense spending that poured money into regions such as Southern California.

America had an expanding economy with prosperity to spare, and could easily afford high union wages and a small welfare state.

With Southern conservatives as a major force in the Democratic Party, Democrats and Republicans were both centrist parties that seemed to have relatively little to fight over – especially since Democrats for more than four decades dominated the Congress.

Republicans occasionally won the presidency. And except for John F. Kennedy and Barack Obama, every other Democrat president since FDR – Missourian Harry Truman, Texan Lyndon Johnson, Georgian Jimmy Carter, and Arkansan Bill Clinton -- has been a Southerner or borderline Southerner.

In the Congress, however, the always-minority Republicans became in the opinion of conservative critics a mere echo of the other side, a "Democrat-lite" party. Republican House Speaker Newt Gingrich later voiced what conservatives thought by describing long-time moderate GOP Senator Bob Dole as "the tax collector for the welfare state."

As the Republican Party became more conservative, starting in the mid-1960s, its candidates began to win in parts of the once-solid Democratic South. It would find a new identity in 1980 as the party of charismatic President Ronald Reagan.

Political Polarization

The Democratic Party at the same time was moving left under the influence of the civil rights, anti-Vietnam War and New Left movements. These Progressive factions would fracture the party in 1968 and put Republican Richard Nixon in the White House.

For several decades, the Gallup polling organization has consistently found that approximately 20 percent of Americans describe themselves as liberal or progressive, and that roughly 41 percent describe themselves as conservatives. We are a center-right nation.

Yet one of our two major political parties, the Democratic Party, has largely shed its Southern conservative roots and become increasingly the party of Big Government and the welfare state. Critics describe it as not so much a genuine political party, but as a collection of special interests that live on taxpayer money – public employee unions and welfare recipients – and government help – such as trial lawyers.

Today's Democrats, say critics, are motivated by only two issues – how can government be made bigger and more powerful? And, will my government check keep coming and get bigger?

It is probably fair to say that if the U.S. Supreme Court tomorrow ruled that those receiving government checks or benefits cannot vote – because to do so would be no different from voting themselves money as Benjamin Franklin warned – the Democratic Party would shrivel to tiny third party status in every election thereafter. (Social Security, to which people paid in, should be exempted.)

Nobody, of course, will prevent government beneficiaries from voting in this land, where 49.1 percent of households have someone living there who is issued a government benefit.

Donkey Drag

This means that American elections have turned into votes between the lambs and the wolves over what – or who – will be eaten for dinner.

Many see the 2012 presidential election as the most important of their lifetimes. It will either re-elect, and thereby validate the policies of, Progressive Democratic President Barack Obama, who favors much higher taxes on the private sector to fund much bigger government, and who has made little secret of his ideological hostility to capitalism, or it will replace Mr. Obama with successful business leader and former Massachusetts Governor Mitt Romney.

Investors will weigh more than who won or lost the November 2012 election. They will pay most attention to what gamblers call the "point spread."

As we discussed in our book *The Inflation Deception*, perhaps 25 percent of our economic growth is affected by what could be called "Donkey Drag," named for the popular symbol of the Democratic Party.

The mere fact that one of a nation's two major political parties is actively anti-business puts a dark cloud over its investment environment.

If voters in 2012 re-hire this party's incumbent president, then investors around the world as well as here will see this as a warning not to put their money at risk in the United States....and potentially to take out what they already have invested. The voters would have endorsed a political party bent on expropriating capitalist money via higher taxes and restrictive regulation.

The Point Spread

Investors will look not only at who wins the 2012 elections, but also at how big the margins between winners and losers are. Here's why:

Even if President Obama or congressional Democrats lose, a close vote might drastically affect how winning Republicans behave thereafter, as follows:

(1) Republicans will be afraid to rock the boat by overturning Obamacare and other Progressive Big Government programs;

(2) Republicans will soften their conservative positions and move to the center, as they have in the past;

(3) Tea Party and other Small Government activists will see Republicans backing off their campaign positions and will have little enthusiasm for again supporting GOP members of Congress in 2014;

(4) Therefore investors will anticipate Democratic congressional victories in 2014 and a potential return to the White House in 2016, and will limit their investment in the United States accordingly.

Whether Democrats win or lose, this is how Donkey Drag pulls down the economy.

In the 2012 election, therefore, only large and sweeping Republican victory margins will produce over the following year a strong, pro-investment environment to revive jobs and the economy.

Anything less than this could leave the economy in an anemic, uncertain condition, with investment money potentially fleeing the country. If 2012's economic malaise continues, it would be used by President Obama in his second term to justify an even larger welfare state and higher taxes.

Win-Win and Lose-Lose

Where Progressives can frighten capitalists out of creating jobs, this then becomes a pretext Progressives can use to fill the vacuum with bigger government. For them, this is a Win-Win, divide-and-conquer strategy for attaining and retaining political power.

It is also a likely Lose-Lose proposition for the economy and America's prospects for future prosperity.

If President Mitt Romney faces such ongoing low growth, high unemployment and a still-Democratic U.S. Senate, then Progressives in the national media will try to blame him and the Republican Party for whatever economic problems are not solved.

If capitalists invest, they risk losing their profits or property the next time Democrats come to power – a cycle that can happen whenever voters get tired of Republicans.

Harvard economic historian Niall Ferguson could have expanded his observation that America is now a 50-50 nation, split down the middle between makers and takers, those who pay the taxes and those who devour them.

Although in their own ways both the Republican and Democratic Parties have supported Big Government, today with the economy and dollar weakened, the two parties are now fighting to grab pieces of a shrinking pie, not the expanding pie of the 1950s.

Part of the stridency of recent politics is born of desperation as the parties must fight more to gain less as the economic pie shrinks for everyone. This is why a pro-growth, small government would be best. It would restore America to economic health so we can again grow the pie for everybody.

"Other People's Money"

Republicans say that they are fighting for the right of taxpayers and producers to keep more of what they earn.

Democrats are now openly the Progressive party of wealth redistribution. They have become America's Social Democratic party, a European-style democratic socialist party bent on redistributing the nation's productivity via coercive politics instead of peaceful, voluntary exchange in the free market.

As former British Prime Minister Margaret Thatcher purportedly said: "Socialism is fine until you run out of other people's money."

The inevitable failure of socialist systems, as we see in today's Europe and yesterday's self-deconstructed Soviet Union, is that they are almost entirely systems of redistribution, not production.

Karl Marx himself wrote that the communism he desired depended on expropriating a capitalist system that had already created enough wealth for communists to redistribute.

Marx himself understood that socialism was not very good at creating wealth. Marx never really answered the next logical question: what does the state do after all the wealth it has looted from capitalists has been devoured?

Parasites

Progressivism is ultimately an ideology of parasites, like tapeworms, mistletoe or leeches. Its system is based on stealing the life energy from producers and innovators.

Nature has many parasitic life forms, yet their survival depends on a fundamental principle: if a parasite sucks too much life energy from its host, the host will die, and so too will the parasite soon thereafter.

A successful parasite feeds on its host but does not take so much that the host weakens and dies.

Some of Europe's Social Democratic ideologues understand this and have tried to find a Third Way between capitalism and socialism that improves the survival prospects of both producers and parasites.

Many Progressives in America's Democratic Party exhibit no such need for balance or intelligence when dealing with the businesses and productive individuals they aim to expropriate.

They are eager to "put their boot on the neck" of businesses – to use a phrase repeatedly used by Obama appointees – and take all they can from the private sector, even if this kills the goose that when left alone can lay golden eggs that benefit everybody.

One interviewer at the 2012 Democratic National Convention asked many delegates and other party faithful if they thought all corporate profits should be made illegal. He was shocked at how many simply answered "Yes."

Contrary to President Obama's belief that government, not businesses, built America, government does not and cannot create prosperity because it makes nothing – except war.

Too much government makes most of us poorer, because everything it redistributes so Progressively must first be taken from somebody else by force or debt or the hidden tax of inflation.

After Free Enterprise

Once the Progressives have expropriated all the fruits of every entrepreneur's labor, and bled the capital out of capitalism, what kind of enterprise do they think will remain after they have killed *free* enterprise?

Only two answers seem likely:

(1) Unfree enterprise, almost inevitably based on government-forced labor, a kind of slavery or serfdom of required government jobs. In the late Soviet Union they had a joke about this: "The Government pretends to pay us, and we pretend to work."

(2) Or no enterprise at all, and hence no more goodies for Progressives to redistribute, as we descend from utopia into a new Dark Age of redistributed poverty for all...except the ruling Progressive elite.

What happens, too, after Progressives propagandize a large segment of society to believe themselves entitled to the wealth of their neighbors?

Most Progressives preach this selfishly – making their government the middleman who expropriates the wealth of the productive "rich" and then transfers it to others, as we noted elsewhere, with government taking a large piece of this wealth as a charge for this service.

Yet once millions feel entitled to their neighbors' wealth, many will eliminate this greedy political middleman and feel justified in directly looting those who have earned the wealth they covet. Progressive class warfare might easily be perceived by criminals as giving them a license to steal.

Growing Government

How well has President Obama done at creating a Progressive utopia?

President Obama's policies have failed to prevent the loss of more than 7 million private sector jobs. He did, however, hire up to 10,000 new federal employees every month, growing the size of the federal workforce by 11.4 percent.

Mr. Obama reportedly is creating 16,000 jobs for IRS agents to monitor Obamacare alone, giving America Progressive health care that its critics say will have the compassion of the IRS and the efficiency of the Post Office.

The average federal employee makes $126,000 each year in wages plus benefits, roughly double the income of the typical private sector employee, according to an investigation by *USA Today*.

This suggests that had President Obama allocated taxpayer stimulus money differently, he could have made possible twice as many new jobs in the private sector as he created in the Federal Government.

President Obama said that thanks to his intervention "General Motors is alive." He carried out a Latin American-style expropriation of auto giants General Motors and Chrysler. In the process he shoved aside secured bondholders and non-union pensions, and effectively gave both companies to the United Auto Workers, a major union contributor to Democratic campaigns, including his. What remained of Chrysler he then transferred to Italian auto maker Fiat.

General Motors has used American taxpayer money constructively. It is, for example, spending $1.5 Billion on new manufacturing facilities – $1 Billion of it in Brazil, and $500,000,000 of it in Mexico.

In August 2012 Mr. Obama in a Colorado speech said: "Now I want to do the same thing with...every industry," in effect carrying out the Old Leftist dream of expropriating private enterprise, America's means of production.

Columnist Ben Shapiro calculated what it would cost taxpayers to acquire every American company thrown into difficulty by the President's economic policies, using what the GM and Chrysler bailouts required. The cost to taxpayers, Shapiro estimated: at least $17 Trillion.

In the long run, any economy able to survive and thrive must produce more than it consumes. For a society to remain free and healthy, individuals should produce more than they consume. An explosive growth of parasites is almost always debilitating and potentially deadly.

Metastasizing government is a parasite that devours the life blood that the productive private sector would otherwise have used to create a more prosperous society. Again, too much government makes most of us poorer.

President Obama has already greatly enlarged the welfare state and the number of people living at taxpayer expense. Nearly one American in seven now gets Food Stamps, an increase of 70 percent in this welfare program since Mr. Obama took office. America now spends nearly a trillion dollars a year on means-tested transfer programs such as welfare – and more on the non-means-tested programs Social Security and Medicare.

Regulations = Taxes

President Obama has not only increased taxes but also costly regulations, another huge burden for which we must pay.

Americans for Tax Reform calculated that in 2011 the average American worked from January 1 until August 12 to reach "Cost of Government Day," the day he or she finished working to pay for taxes, along with often-hidden regulation and mandate costs, and could only then start working for themselves and their families.

This means that in 2011 the average American worked 224 days of the year for the government – more than 61 percent of his or her work time – as a de facto government employee.

"Contributing to the coming crash of 2013 is Obama's regulatory storm," wrote economist Stephen Moore, author of *The End of Prosperity: How Higher Taxes Will Doom the Economy—If We Let It Happen*, in the October 2012 *Newsmax* Magazine.

"Academic studies estimate the total economic costs of regulation to be rapidly rising toward $2 Trillion per year, or $8,000 per employee. That is close to 10 times the corporate income tax burden and double the individual income tax," wrote Moore.

"Those costs will effectively be yet another major tax increase on the economy of trillions over the years," Moore continued. "Particularly devastated will be energy-intensive manufacturing....Delayed until after the election are the EPA's [Environmental Protection Agency's] proposed ozone rules, which can prohibit business expansion in up to 85 percent of the country that would be found in violation."

All government regulations and mandates are, in effect, additional taxes that expropriate pieces of your life, liberty and income.

Regulations also give politicians and bureaucrats the means to punish or shake down people and companies, and by selective enforcement to pick winners and losers in the economy.

In 2012, after the chief executive of a chicken fast-food company told an interviewer that he personally did not approve of gay marriage, the Progressive mayors of two major cities rushed before television cameras to say they would not allow him to open a franchise in their communities. How? Apparently by capriciously denying his company required permits and approvals.

Frankly, why would any business person invest in a city where people have elected a Mayor who uses Political Correctness and abuses the law to decide who is permitted (literally) to open a business there?

The Welfare-Jobs Con

As we noted earlier, in 2012 the Obama Administration altered President Bill Clinton's successful bipartisan 1996 reform that had reduced welfare rolls by up to 50 percent.

The Obama Administration allowed states to eliminate the requirement that able-bodied people on one of the biggest welfare programs had to demonstrate that they were actively looking for a job.

Without this Clinton reform, the social safety net can easily turn into a hammock of free money, free housing and many other benefits.

By giving states and localities a chance to get federal waivers from this work requirement, President Obama opened the way for millions to re-apply for welfare – and for dependency on Big Government that could give them a selfish reason to vote for America's Big Government Party.

Yet this cynical, and probably illegal, change to the Clinton welfare reform could also have boosted Mr. Obama's re-election chances in a second way.

When welfare recipients were required to seek work, they were counted as unemployed.

If the Obama waivers allow millions on welfare to stop seeking jobs, then within four weeks they would no longer be counted on the unemployment rolls.

This apparently was designed to have the potential, in as few as two months, to produce a drop in the national unemployment rate, which the President prior to the election could then point to as evidence that his economic policies were succeeding.

The only thing that would have changed, however, is that millions of welfare recipients were no longer seeking work.

It's not good news for taxpayers that President Obama tried to buy his re-election with tens of billions of taxpayer dollars for the millions of people his new policy could potentially add to the welfare rolls.

All these billions of dollars must either be borrowed from entities such as the People's Republic of China, our biggest foreign lender, or printed out of nothing, which causes inflation and raises the price of everything working Americans must buy. Either of these options is regressive, hitting hard the pocketbooks of the poor.

Targeting the Rich

Or Progressive taxes can be massively raised on "millionaires and billionaires." This may be cynically shrewd politics, but it is lousy economics.

More than two-thirds of America's wealthy are also the nation's stock and bond holders and company owners. Money taxed from them is not "wealth." It is "capital," as in capitalism – the seed corn that, if government does not devour it, would go into expanding businesses and hiring workers.

Just as Adolf Hitler created propaganda demonizing Jews to justify his expropriation of their property, class warfare Progressives demonize the rich to justify the taking of their property.

Progressives know perfectly well that most of their added taxes on business – which in America are already the world's highest among advanced economies – will simply be passed on in the form of higher prices and be paid by the poor.

Progressive politicians then denounce corporations for their high prices, never mentioning how much of those prices are actually passed-on taxes.

Politically the Progressive game is brilliant – make capitalists pay through the nose, then use businesses to do the dirty work of extracting taxes from everybody via higher prices so the politicians are not seen as big taxers.

Passed-on Taxes

President Ronald Reagan had economists look at a staple of life so basic that most states exempt it from sales tax – a loaf of bread.

The price tag on a simple loaf of bread, they found, conceals at least 151 hidden taxes – the property tax of the farmer, the road use and fuel taxes paid by the trucker, the Social Security taxes of the Supermarket, and so forth.

At least 151 such taxes are passed on to you in the higher price of bread – and pretty much everything else you buy. They are a burden that every producer and every buyer – even those deceived into believing that they pay no taxes –carries.

Progressives then use these high prices to declare that a family of four making less than $21,000 a year is in "poverty." Government now spends $61,000 each year for every American in poverty, yet up to 80 cents of every dollar to help the poor is devoured by the government itself and never reaches the poor.

Never Enough

To provide an always-increasing flow of money to pay government union workers, welfare recipients and welfare state employees is difficult. Politicians cannot raise taxes to infinity – especially during the economic downturns that their Keynesian stimulus policies never really prevented.

The chief source of such ever-increasing money has become inflation – simply printing more and more fiat paper, and letting the citizenry pay for the new economic order through the hidden tax of ever-rising prices.

Simply put, the very existence of a welfare state – and of the political party elected by welfare beneficiaries – depend on inflation, on the power of government to keep welfare money coming and growing, whether the economy rains or shines.

This can be guaranteed only by politicians willing and able to print however much debased currency they need to buy the votes to stay in power.

Alan Greenspan, an economist who would later be chairman of the Federal Reserve, explained in 1966 why the Progressives running our government had to turn the U.S. Dollar into a paper fiat currency. We examine his explanation later in this book.

This means that a symbiotic nexus exists between inflation and the modern welfare state, which could never have expanded had Progressives not radically transformed our currency in 1913.

Perhaps this is why British author Sylvia Townsend Warner wrote that "Inflation is the senility of democracies."

Implicit in Keynesianism is the unspoken idea that humans now have the power to play God, to repeal the Law of Supply and Demand and have a few superior unelected technocrats manipulate the entire economy in what they decide is our best interest.

Keynes and his disciples were elitists who deemed their rule over the economy superior to the free market ideal of separation of economy and state.

Today we have seen President Obama's Keynesian stimulus efforts produce anemic growth and longtime high unemployment, just like a European welfare state, despite the injection into the economy of trillions and trillions of dollars.

Today serious economists outside the Progressive-left have become skeptical of Keynesian assumptions. One problem they now recognize is that stimulus might provide benefits, but it always imposes costs on an economy.

When stimulus money comes from taxes, government has taken it away from investors and companies that can no longer use it to expand and hire – and that probably would have spent it more productively than government.

When stimulus money comes from borrowing, taxpayers must repay it with interest. Government borrowing during the Great Recession has also "crowded out" many businesses that were trying to get loans from a limited lending pool.

When stimulus money is created by the Fed or government simply printing money out of thin air, this devalues the dollars from honest productivity that people have earned and saved, a debasement of our money we experience as inflation. It is a cruel, regressive form of taxation that hits the poor and those on fixed incomes hardest.

Simply printing money from nothing as the Fed and government do, truth be told, is a form of counterfeiting that anyone else would go to prison for doing.

Bottom line: Keynesian stimulus is not free. It is painfully expensive.

No Free Lunch

As the conservative-libertarian science fiction writer Robert A. Heinlein famously said: TANSTAAFL. This is an acronym for the truism "There Ain't No Such Thing As A Free Lunch."

You might be invited to an event where ham sandwiches are being given away, but somebody paid for them. Somebody grew the wheat and baked the bread, raised the pigs and milled the mustard, and paid for the gas to drive the completed sandwiches to this event.

Once upon a time, almost all of us were producers. We understood by our calloused hands and the sweat of our brow the real cost of making what we consumed. We were in the world as makers, not just takers.

Today we live in a surreal American economy where more than 70 percent of our Gross Domestic Product comes not from production but from consumption. Many have grown fat, thanks to an obese government's largesse.

Many now understand the European consciousness that the 19th Century French political philosopher Friedrich Bastiat was mocking when he wrote:

"The State is that great fiction by which everyone endeavors to live at the expense of everyone else."

Bastiat then added: "They forget that the State lives at the expense of everyone."

Annexing the Economy

By creating a huge new welfare dependency and debt, and the illusion of improving employment to improve its candidates' chances for re-election, the Big Government Party makes companies even more hesitant and apprehensive about hiring and investing in America.

This fear leads to even fewer jobs for Americans who want to work, and more people becoming dependents of the government.

"When governments annex a huge chunk of the economy, they also annex a huge chunk of individual liberty," writes author Mark Steyn.

"You fundamentally change the relationship between the citizen and the state into something closer to that of junkie and pusher – and you make it very difficult ever to change back."

"In the end, it's not about money, but about something more fundamental," writes Steyn.

"Yes, you can tax people to the hilt and give them 'free' health care and 'free' homes and 'free' food. But in doing so you turn them into, if not (yet) slaves, then pets. And that's the nub of it," writes Steyn. "Big Government leads to small liberty, and to small men."

"The opposite of Big Government," writes Steyn, "is not small government, but Big Liberty."

Or as Paul Ryan, a Congressman from the very Progressive state of Wisconsin and Republican Mitt Romney's 2012 vice-presidential running mate, has said, Big Government creates a world "in which everything is free – except us."

Chapter Seven
The Mandate of Heaven

Tipping Point # 2
America's Weather-Wealth Disappears

"And I called for a drought upon the land,
and upon the mountains, and upon the corn,
and upon the new wine, and upon the oil,
and upon that which the ground bringeth forth,
and upon men, and upon cattle,
and upon all the labor of the hands."

– The Bible, Haggai 1:11

An old proverb tells how a kingdom was lost because a battle was lost because a rider was lost because his horse was lost because a horseshoe was lost – "all for the want of a horseshoe nail."

The U.S. Dollar depends on the economy, which depends in part on American food production, which depends on what could be changing weather and climate. This is one factor among many, yet if severe droughts and floods, heat waves and deep freezes, begin coming more frequently, this could be the straw that breaks the economy's, and therefore the dollar's, back.

Cooked Fish

During August 2012, in a scene like the 10 Plagues of ancient Egypt, thousands of dead sturgeon, catfish and carp washed up on the banks of the Platte River in Nebraska.

With the Midwest baking in record-breaking heat and the worst drought in half a century, the river water became too hot to retain the oxygen these fish needed.

In a sense, fish were thus “cooked” to death in several heated Midwestern rivers and lakes.

In one Illinois lake, a flood of dead and dying bass and catfish clogged a power plant’s cooling water intake screen, forcing one of its generators to shut down.

The Mississippi River’s flow shrank to near-record lows, constricting and slowing barge traffic while Army Corps of Engineers dredges worked frantically to clear channels in the shallow river bottom. On one day the shrinking river stranded 93 barges.

At high tide, salty water from the Gulf of Mexico surged many miles up the Mississippi from the Gulf of Mexico, putting the drinking water supply of several towns at risk.

Painted Lawns

In Missouri, lawns withered. Many homeowners responded by painting their dead brown grass green.

In their small backyard Ryan and Peggy had planted a garden – a few tomato plants, beans, squash, melons and corn, as they during childhood had watched their grandparents do – hoping to save on the local supermarket’s rising prices.

Their hopeful garden endured, as they did despite their underwater mortgage. Yet the garden’s plants drooped and wilted in the heat unless revived with water at least once or twice every day.

Their struggling plants produced little food. And when their water bill arrived, Ryan and Peggy were shocked to discover that trying to grow their own food was now more expensive than the supermarket.

In today's hard-pressed American dream, their home garden was no longer Eden.

A Shriveled Heartland

Their garden was a tiny microcosm of the widespread devastation that the Great Drought that began in 2012 brought to America's crops.

By August 2012, that summer's drought affected more than 61 percent of the continental United States and 88 percent of the cropland where corn is grown. More than 1,000 counties had been declared natural disaster areas.

In the western states, another crop was in peril as fires ravaged tinder-dry forests, destroying their harvest of future lumber, and sent exurban homes up in flames.

Redistribution of the Weather

Was this devastating heat and drought a sign of global warming? Not necessarily. Global warming would likely bring more, and more destructive, hurricanes – yet the United States has in recent years been hit by fewer hurricanes than usual.

The drought may have been triggered by a transition of cyclic weather phenomena known as a La Niña to its opposite, an El Niño, both changes in Pacific Ocean regional surface temperatures that can redistribute global rainfall and cause odd bends in world-circling winds, including the Jet Stream.

In 2012 the Jet Stream moved much farther north than usual over North America, bringing hot, dry air northward with it. This drastically reduced rainfall from Texas to the Canadian Maritimes.

(It also may be what reduced the number of 2012 tornadoes by eliminating most of the seasonal collisions between warm southern and cool northern air masses that spawn these high-speed swirling twisters.)

When the Jet Stream zigs far north in one part of the world, it usually zags far south in another. Across the Pacific, killer torrential rainfall flooded southern Japan. Beijing, China suffered the heaviest rainfall in 60 years.

Across the Atlantic, Great Britain in 2012 had the heaviest June rainfall ever recorded. Johannesburg, South Africa – as far south of Earth's Equator as Miami is north of it – awoke to a rare snowfall.

Weather Wealth

The 2012 drought may portend something far more ominous.

It could mark a return to "normal" weather and climate in America's heartland.

Those of us who came of age during the 20th Century tend to think of its weather as normal, yet the last century was the balmiest in perhaps 1,000 years in North America.

Recent weather (as well as that 30-year averaging of weather that defines climate) has been remarkably pleasant compared to the more typical weather extremes of the 18th or 19th Centuries.

Good weather and climate are a natural resource, just like oil or coal or timber.

Nations blessed with abundant good weather can grow food reliably at relatively low cost. They need not spend a fortune to irrigate crops, or huge amounts on energy to avoid freezing in winter or baking in summer.

The United States built its independence and wealth on a foundation of relatively good weather and climate during the 20th Century, as we grew four times more food than we consumed and became the breadbasket to a hungry world.

America prospered from an abundance of good weather. One could almost say that the United States in the 20th Century was the Saudi Arabia of global food because we had ample rain and rich topsoil.

What happens to us if the rains stop coming?

Welcome Back to "Normal"

So what is "normal" weather in North America?

The Great Blizzard of 1888 shut down America's Northeast. It piled snow more than 30 feet deep in Manhattan and forced many residents to enter and exit their homes via third-floor windows.

The 1880s brought years of "Blue Snows" in the northern Great Plains. In springtime, farmers reported finding carcasses of cattle high up in trees after the poor animals froze to death trying to escape through snowdrifts up to 100 feet deep.

In 1815-16 the Northeast shivered through "the Year without Summer," during which snow remained year-round in the shadows of rocks and trees.

The Northern Hemisphere had been chilled by a shroud of stratospheric smoke from two major volcanic eruptions that dimmed sunlight and caused year-round chill around the world.

During America's Revolutionary War, British soldiers dragged heavy cast-iron cannons across the thick ice between Staten Island and Brooklyn. Seaports from Boston to Baltimore then often froze shut during wintertime.

The "Great American Desert"

During the mid-1800s a steady stream of pioneers in covered wagons traveled west on the Oregon Trail.

What today we think of as cropland was called by them the "Great American Desert."

From Nebraska westward, they saw not 10-foot-tall cornstalks but occasional cactus in a fragile, almost-treeless prairie of Buffalo Grass sod where drying winds blew relentlessly, and rainfall averaged less than 20 inches per year.

The soil in these landscapes was rich and deep, deposited by the glaciers and melt-off floods of hundreds of thousands of years of successive ice ages, according to scientists. At the last ice age's peak nearly 20,000 years ago, ice thousands of feet thick lay atop what today is Chicago.

Finding vast expanses of the world's richest topsoil, settlers soon shaped homes from squares of cut sod. These sodbusters eagerly deep-plowed the Buffalo Grass under to open plantable farmland.

And for a few decades they and the nation prospered from unusually bountiful rains. It seemed that the land sellers and government agents were right, that "rain follows the plow."

The Dust Bowl

In the 1930s, however, the rains stopped coming. During a single decade, four different droughts hit 46 of America's then-48 states, turning the central plains into what documentarian Ken Burns calls "an American Sahara."

"The government is one of the reasons the Dust Bowl happened," says Burns, whose television documentary "The Dust Bowl" was scheduled to air in mid-November 2012 on PBS, the Public Broadcasting Service.

"The government encouraged the settlement of this marginal land," he says, and "encouraged the land companies and the railroads to sell this land."

"The Homestead Act had been increased, so in some ways the government was responsible for tearing up this thing."

Black Blizzards

Where farmers had plowed under the protective Buffalo Grass, hot dry winds now sandblasted the fragile soil, blowing it up into vast "black blizzards" that darkened skies as far away as Washington, D.C.

Those who stayed, expecting good weather soon to return, saw their cattle, and in some cases their children, sickened and killed by dust-caused pneumonia. During storms, the air was almost unbreathable.

Many who fled their dust-choked, ruined farms and shattered dreams migrated west, becoming the crop-picking Okies and Arkies that John Steinbeck wrote about in his famed novel *The Grapes of Wrath.*

As the Dust Bowl droughts eased, the remaining farmers began planting rows of trees as windbreaks, and adopting contour plowing and other methods to reduce their susceptibility to this nightmare's return.

A Mountain of Millet

Most of all, farmers in the plains found an alternative source of irrigation water that eased their dependence on unreliable rainfall.

This source is the Ogallala Aquifer, a vast underground deposit of water that stretches from the southern edge of South Dakota to northern Texas. It became the key that unlocked America's immense grain-growing potential.

Ryan experienced that grain potential as a child. Visiting the factory where his father worked, he stumbled upon a warehouse full of grain, a mountain of millet that he climbed, marveling as his small feet sank into the enormous pile of light-brown roundish seeds.

Back then the government bought up millions of tons of surplus grain, which provided farmers with a customer and some measure of price security.

The government, Ryan later learned, usually rented spare warehouses from all sorts of companies, stored the grain for a few years, and then put it on ships and dumped it at sea.

Even then, this bizarre government policy seemed crazy to Ryan in a world where his parents told him to clean his dinner plate because children were starving in China.

Double-edged Plowshares

The U.S. also gave away vast amounts of such grain around the world, but this often proved to be beating double-edged swords into double-edged plowshares.

When an American shipload of free food arrived in a poor country, it immediately destroyed the income of local farmers who could not sell what they grew in competition with free grain.

As a result, local farmers abandoned their fields and moved to overcrowded cities to join the others there who lined up for free American food. This left such nations even less able to feed themselves, and even more dependent on our free food.

American aid was often devastating and destabilizing to local Third World economies, yet as with welfare at home it gave our government a position of power over those who came to depend on our largesse.

Even without intending it, American food and our abundance became weapons because our plowshares had the power of swords.

The United States has striven to teach other nations better ways to produce food. Some more cynical foreign leaders, if blessed with our position as the world's "Saudi Arabia of food," would probably feel tempted to use this to advance their own power, influence and wealth. Power accrues inevitably in those nations with abundant energy or food they can provide to others.

Fossil Water

What we have learned is that the Ogallala Aquifer, on which so much of our grain abundance depends, is shrinking as we and our crops drink from it.

Like the topsoil of America's heartland, this aquifer is also a deposit built up during past ice ages.

Think of it as "fossil" water. When we have pumped this aquifer dry, it may take as much as 1,100 years to refill naturally (although recent research suggests that parts of it may refill faster than this).

As to all that grain we gave away or dumped at sea, think of a ton of grain as being made up of a ton of topsoil.

When that "fossil soil" is gone, it might not renew itself for millions of years. And now we are turning our topsoil into corn, and our corn into an inefficient substitute for oil, as we shall soon discuss.

An Empire Built on Weather

America's wealth and power have largely been an empire built on weather, past and present.

That weather may now be changing, returning to normal, and our deposits of water and topsoil may be more temporary than we understood.

Our key crops may depend on natural deposits of water and soil gathered for us by past ice ages.

The good growing weather of the 20th Century may have been a temporary once-every-1,000-years cyclic episode of balminess and adequate rainfall in a Midwest that 85 percent of each Millennium is a drought-prone quasi-desert unsuited, without irrigation, to growing crops such as corn or wheat.

If so, then we can probably continue to produce such crops. The cost of doing so will rise, however, as we must spend much more on importing irrigation water and using more fertilizers and pesticides to offset worsening weather.

America's crop heartland might soon have to import water from hundreds of miles away, as Southern California does, which is an expensive proposition that would greatly increase food prices. Some coastal Southern California cities are preparing to desalinate sea water, also a costly way to get usable water.

The 10 percent surge in world food prices associated with bad weather here, in India and in Russia during the summer of 2012, will increase as the drought's longer-term effects ripple through the economy in late 2012 and throughout 2013, jolting prices for high-fructose corn syrup, ethanol, and many other products.

How severe was the drought we experienced during the summer of 2012? It was the worst in more than 50 years, and it might not be over.

The Dust Bowl, as we noted, was actually four different droughts that piled up cumulatively in a single 10-year period. The 2012 drought in America was more severe than any one of the Dust Bowl droughts.

Field of Dreams

Despite these problems, two things continued to grow and thrive in the 2012 drought's baking heat.

Marijuana plants, known in America's heartland as "Iowa ditchweed," seemed unfazed while corn and other crops withered and died around them.

The business cable channel CNBC reported that this gave the movie "'Field of Dreams' a whole new meaning."

The other thing that kept growing was government, noted Dan Holler of Heritage Action for America, a project of the Heritage Foundation.

"Don't Waste the Drought," wrote Charles Fishman, author of *The Big Thirst: The Secret Life and Turbulent Future of Water*, in the August 17, 2012, *New York Times*. He was echoing Chicago Mayor and former Obama Administration enforcer Rahm Emanuel's famous statement: "You never let a serious crisis go to waste....it's an opportunity to do things you think you could not do before."

Past crises and government programs have expanded a government bureaucracy so much that the U.S. Department of Agriculture now has one employee for every 11.4 actual American farmers.

The Food Stamp program began in an unholy alliance between members of Congress from farm states, eager to justify more government crop subsidies and higher price supports, and those from big cities, eager to have free goodies they could give away to buy votes.

In the 2012 drought, congressional Democrats used this crisis to massively lard the pending Farm Bill, but not with much of the kind of help for endangered farmers that in the past had turned Iowa into a welfare state. The new legislation added tons more money for urban Food Stamp recipients, who stand to devour between 70 and 80 percent of all money in a bill nominally funded to help farmers.

Democrats then blocked a Republican measure providing emergency aid for drought-stricken farmers, holding the farmers hostage to liberal demands that the Farm Bill, more than 70 percent of whose spending was for more big city freebies, be passed instead.

The Long Soup Line

Under the economic collapse of the Obama Administration, the number of Americans on Food Stamps (renamed SNAP, the Supplemental Nutrition Assistance Program) has grown by 50 percent to 46.37 million.

This growth of a welfare dependency program has come partly through aggressive government advertising, including the funding of Spanish-language soap operas, telenovelas.

Food stamp use has also grown because Obamanomics has driven the nation into what for millions of low income families seems like a Great Depression.

Most Americans do not see this because Depression-era soup lines have been replaced by a welfare check or food card in the mail.

If today's more-than-46 million food stamp recipients were gathered into a Progressive Soup Line, with each person taking up only one foot, this line could stretch from New York City south to Atlanta, then west to Los Angeles, then north to Seattle and east again to New York City, encircling most of the nation.

Enough food stamp recipients would then be left over to create a second line that stretched from Los Angeles east to Chicago and then from Chicago south to Houston. If each person took up two feet, the Obama Depression soup line would stretch 17,564 miles, more than 70 percent of the distance around our planet.

Such an immense soup line would be evidence of Mr. Obama's Invisible Depression, which is why you will never seen it illustrated this way on the mainstream media.

Eco-Mirage

Approximately 40 percent of America's corn harvest now goes to make ethanol, an alcohol that the Federal Government since 2005 has mandated be blended with gasoline to reduce greenhouse gas emissions.
"The drought lays bare the folly of trying to expand an industry where the economic fundamentals don't make much sense," wrote the editors of

Bloomberg News of this ethanol mandate in August 2012.

What is wrong with clearing the air and creating a domestic fuel extender to ease our dependence on foreign oil? *Bloomberg*'s editors counted some of the ways this government mandate makes no sense:

If ethanol magically appeared out of thin air, then blending it with gasoline under ideal circumstances might reduce pollution by as much as 20 percent. The trouble is, it does not appear magically.

Ethanol comes from corn that requires vast amounts of scarce water, precious cropland, and oil-based chemical fertilizers and pesticides.

This corn must be planted in plowed fields, irrigated during dry times, then harvested and processed by energy-gulping machines.

When these costs are included, ethanol consumes far more energy than it saves and is more environmentally polluting than gasoline.

The resulting ethanol in energy content is around 50 percent more expensive than gasoline, and it comes with other high price tags as well.

Corn Liquor

Diverting corn and cropland to make ethanol causes Americans to pay at least an extra $40 Billion every year in higher food costs, according to researchers at Texas A&M University.

Ethanol is so corrosive, noted *Bloomberg News*, that it cannot be shipped via pipeline – but only in lined tanks via railroad cars or trucks, both of which are less efficient than pipelines and increase the environmental and financial cost of using it.

Ethanol is corrosive in your automobile as well. It can damage the seals of car engines if blended to be as little as 4 percent in a gasoline mix.

Such damage reduces engine life, causes more breakdowns, and increases pollution. Despite this, the Obama Administration's regulators have required that this ratio of ethanol in some gasoline blends be exceeded.

When Government Runs Markets

Ethanol is also a direct burden to taxpayers, who according to the Congressional Budget Office are required to pay $1.78 in subsidies for every gallon of gasoline that corn-based ethanol replaces.

The giant companies that receive these subsidy payments have been remarkably generous to lawmakers, providing everything from campaign contributions to travel on corporate jets with the money you pay in taxes.

What do taxpayers get in exchange for this subsidy? More pollution. Higher food prices. And with the price of corn driven up more than 50 percent by the drought, the coming price jump in ethanol-added fuel will also siphon lots more dollars from driver wallets at the gas pump.

We are left with one more very wasteful government program – and with the absurdity that the government will ticket you for driving while impaired by alcohol, yet the government now requires that our cars guzzle corn liquor in amounts that impair their performance and safety.

In January, 2012, the *New York Times* reported that the oil companies are being required to pay $6.8 Million in government fines each year "for not using a biofuel that doesn't exist," failing to include in their fuel blends cellulosic fuel that no company is producing in such commercial quantities. When asked about the unfairness of this requirement, an Obama Administration official reportedly replied that the oil companies could afford to pay the fine. It appears to be just one more rule used to squeeze money out of the private sector.

A Hungry World

Ethanol's highest cost, however, goes unseen by most Americans. It happens in Third World nations where America's ethanol mandate forces up the price of tortillas and other corn-based foods on which the poor depend.

Mr. Obama's cold-hearted refusal to suspend the ethanol mandate during the 2012 drought has unquestionably produced malnutrition and deaths in some of the world's poorest villages and families.

Most will die quietly. The rising cost of bread in Tunisia, however, led to rioting and the overthrow of its government in 2011. Rising food costs were a factor in the popular uprising that overthrew a government friendly to the West in Egypt.

Droughts and foolish government policies throughout history have produced economic and political instability and triggered huge consequences, including the French Revolution.

Hungry people tend not to make good decisions. Your mother was right: a full belly can produce a smarter brain.

We will never know how many thousands have suffered and died because of President Obama's toxic blend of crony capitalist subsidies for his political contributors and his extreme environmental ideology. We may soon feel yet more damaging and costly "blowback" from his policies.

The Mandate of Heaven

When drought and other weather disasters struck ancient China, some people took this as evidence that their Emperor had lost divine favor, that he had "lost the Mandate of Heaven" and should be replaced to restore harmony between heaven and earth, between humankind and nature, and to bring back divine blessing to the land.

The Bible likewise comprehends a link between virtue and the blessing of good weather:

"If you walk in my statutes
and observe my commandments and do them,
then I will give you your rains...
and the land shall yield its increase....
And you shall eat your bread to the full
and dwell in your land securely."

– Leviticus 26:3-5

In July 2012 Secretary of Agriculture Tom Vilsack told reporters that he was praying for rain.

"I get on my knees every day, and I'm saying an extra prayer right now," he told reporters. "If I had a rain prayer or a rain dance I could do, I would do it."

The atheist Council for Secular Humanism criticized Secretary Vilsack for saying he prays for rain to end the drought.

"There are two ways
to conquer and enslave a nation.
One is by the sword,
the other is by debt."

– John Adams

Chapter Eight
The Dollar Engulfed

Tipping Point # 3
The Military-Money Synergy Collapses

"The single biggest threat
to our national security
is our debt."

– Admiral Mike Mullen
Former Chairman
Joint Chiefs of Staff

American success in the global economy has stood tall on two mighty legs – one military, the other economic. These legs have worked together synergistically, with each strengthening and reinforcing the other.

America's military might has given us the credibility and security to prosper in global trade and to keep sea lanes and other supply lines open.

America's economic strength has given us both wealth and the industrial and technological capability to create and fund the most powerful military on Earth.

If either leg weakens, however, we may be unable to remain standing tall for very long in today's unstable world.

The Great Debasement has certainly weakened America's economy and dollar, but it has done worse than make us poorer.

It has increased our risk of war, including nuclear war, in the immediate future.

The Great Debasement is costing us more than money. It could cost Americans, as well as our children and grandchildren, their lives through terrorism and war.

Debt Threat

America's trillion-dollar budget deficit "poses a national security threat in two ways," warned Secretary of State Hillary Clinton in a September 2010 speech. "[I]t undermines our capacity to act in our own interest, and it does constrain us where constraint may be undesirable."

America's soaring debt "sends a message of weakness internationally," said Secretary Clinton.

Ms. Clinton said it was "personally painful" to see such a deficit after her husband, President Bill Clinton, had ended his second term in office with projected budget surpluses.

She did not mention that President Clinton took hundreds of billions of dollars from America's defense budget to bankroll more social spending, just as President Obama has. This has made America and its forces more vulnerable to terrorist and other enemy attacks.

"American economic weakness undercuts U.S. leadership abroad. Other countries sense our weakness and wonder about our purported decline," warned Brookings Institution foreign policy scholars Kenneth Lieberthal and Michael O'Hanlon, two authors of the book *Bending History: Barack Obama's Foreign Policy*.

Perception of Weakness

"If this perception becomes more widespread, and the case that we are in decline becomes more persuasive," they wrote in July 2012, "countries will begin to take actions that reflect their skepticism about America's future."

"Allies and friends will doubt our commitment and may pursue nuclear weapons for their own security, for example; adversaries will sense opportunity and be less restrained in throwing around their weight in their own neighborhoods," they say.

"The crucial Persian Gulf and Western Pacific regions will likely become less stable. Major war will become more likely."

"[F]or Obama or his successor," Lieberthal and O'Hanlon conclude, "there is now a much more urgent big-picture issue: restoring U.S. economic strength. Nothing else is really possible if that fundamental prerequisite to effective foreign policy is not reestablished."

War Game

Patrick sat quietly, looking older as candlelight painted flickering shadows on his face.

"War games are not played for fun," he said at last, "but to appreciate why experts feel urgent fear and foreboding about America's weakening economy, let's lay out a fictional scenario or two about what the huge transfer of wealth from America and the West to other lands might produce."

"Lieberthal and O'Hanlon fear war in the Persian Gulf from America's economic weakening, so let's play a War Game," said Patrick.

"This body of water has watched great empires rise and fall – Babylon, the Tower of Babel, Baghdad, Alexander the Great – over many thousands of years. It is still watching."

"Imagine a hot June day on the Persian Gulf in 2013 A.D.," said Patrick. "The 100-degree temperature conjures a mirage-like shimmer of heat waves above the Gulf waters."

The Ship-devouring Wind

The Shamal wind is blowing from the northern desert, carrying dust and charged ions that alter brain chemistry and make people feel a little crazy, explains Patrick.

For thousands of years, Gulf fishermen have feared this wind they call *Barih Thorayya*, after the constellation we call the Pleiades and the Japanese call *Subaru* – a violent wind which they believe devours ships.

An American fleet of warships, including two aircraft carriers, are confined in the narrow waters of the Persian Gulf.

America's armada is in these foreign waters for one reason: most of the nations around this body of water possess vast reserves of oil, which at this moment in history is the lifeblood of the developed world.

This hydrocarbon goo had little value two centuries ago and may have relatively little value two centuries hence, yet by a convergence of geography and technology, in our time the Persian Gulf is the world's most valuable exporting source of the liquid petroleum many call black gold.

The United States could already be the Saudi Arabia of other precious hydrocarbons useful as fuel – with an 800-year supply of coal and apparently limitless deposits of clean-burning natural gas.

The U.S. also produces 7.8 million barrels of oil each day, only a million fewer than Saudi Arabia. The problem is that we consume 18.8 million barrels every day, a costly addiction when oil prices get anywhere near $100 per barrel.

The world is rapidly acquiring better means to find and extract such riches, and as more nations develop their own wells, the world's dependence on Persian Gulf oil has begun to decline.

Israel, for example, has found vast natural gas deposits beneath the Mediterranean near its coast.

Ireland may soon be pumping oil each year worth half its annual Gross Domestic Product from an undersea deposit 40 miles off its coast.

And yet our Navy polices the Persian Gulf, a force to stabilize the world supply of oil for ourselves and our allies. This is a risk and an expense that the United States has long deemed essential to safeguard the world economy.

Oiling the World

The Persian Gulf is roughly 600 miles long and, on average, only 150 miles wide, a distance that some modern classified anti-ship weapons can traverse within a few heartbeats or less.

The U.S. Navy is magnificent but not omniscient. In August 2012 in the Persian Gulf, the U.S.S. Porter, a Navy guided missile destroyer, accidentally crashed into the Japanese-owned oil tanker M/V Otowasan. In the Gulf of Mexico, a Russian submarine reportedly recently cruised around for several days before the U.S. Navy detected its presence.

Much of the Persian Gulf – or Arabian Gulf, depending on who your nation's allies are – is relatively shallow, especially near its northern end where some archeologists believe the now-submerged Biblical Eden and its convergence of four rivers once existed.

At the Gulf's southern end is its passage to the Indian Ocean, a choke point only 21 nautical miles wide called the Strait of Hormuz. It is the "jugular vein" of the developed world's oil-dependent economy.

One-fifth of the world's shipped oil passes through this strait under the noses of Iranian anti-ship and anti-aircraft missiles, weapons capable of slashing open this vital vein within seconds.

On this quiet day, a third U.S. aircraft carrier has sailed through the strait, presumably to replace one of the two carrier groups already there.

Nearly 60,000 feet overhead, what appears to be an ordinary weather balloon has slowly drifted westward from Iraqi airspace out over the western shore of the Persian Gulf. The Navy has tracked this balloon since first noticing it over Iraq, yet sees no reason for concern.

In fact, this balloon was quietly launched from Syria into the Shamal wind by operatives of the Iran-backed Shiite "Party of God," Hezbollah.

A Second Sun

In an instant, the sky above the western Persian Gulf acquires a second sun as a compact multi-megaton nuclear weapon, sheathed in stealth materials and disguised as weather-sensor equipment beneath the balloon, explodes. Nine miles high, the flash blinds several sailors who happen to be looking in its direction. Several carrier aircraft out on missions are sent crashing into the Gulf.

The blast's shockwave does not destroy America's three aircraft carriers, but the Electromagnetic Pulse (EMP) from the blast blinds a wide array of American military electronics.

Their optical fiber systems survive the pulse, but older military computers "hardened" against EMP by using gallium arsenide instead of silicon chips instantly fry, both in the carrier groups and at American military facilities in Iraq, Saudi Arabia and Qatar.

In Bahrain the EMP destroys this global financial center's banking computers and data bases that shepherd much of the world's unseen juggling of the Euro. Within seconds, banks and exchanges across Europe begin to fall like dominoes, sucking the world economy into a vortex of chaos.

Barrage

In that moment of American military disorientation, Iran launches thousands of assorted weapons – missiles, laser cannons, torpedoes, aircraft and more – at the American carrier groups.

The Navy's Metalstorm and several other far more secret defensive systems respond, stopping most of what is incoming. But the EMP attack has impaired enough electronic systems that some of the Iranian weapons find their target.

One carrier becomes a radioactive cinder under the mushroom cloud of an atomic-tipped missile. A second is hit and burning, at risk of explosion.

Thousands of American sailors are dead and dying. What commands does the Commander-in-Chief have for our forces? How shall we retaliate?

The President hesitates. Surely this is a misunderstanding, he tells aides. Is our intelligence certain that this was Iran? Why, he asks an aide, would Iran do this?

Fire and Sword

"Mr. President," the aide soberly replies, "the rulers of Iran believe that a nuclear war with Israel and the United States will bring their 12th Imam, their messianic Mahdi spoken of in the Hadith as destined to win the final global triumph of Islam by fire and sword."

All three aircraft carriers have now been destroyed, each with a crew of nearly 5,000.

"They're either at your feet or at your throat," a Middle East expert had warned the President privately. "They respect only one thing, and that is strength. If you show any kind of weakness, military or economic, they will view you as vulnerable and attack. That's why they destroyed the World Trade Center."

The President orders no retaliation against Iran for the attack, ascribing it to unidentified attackers.

Liberty Falls

"In war-gaming, security analysts look at risks that are likely, yet would do relatively little harm. And they also consider risks that are unlikely but would have terrible human and financial consequences if they happen," said Patrick. "In a world where North Korea, Pakistan and very soon Iran have nuclear weapons, think tank futurists have considered what would happen if, as their 'delivery system,' they used not a missile but a terrorist with a minivan to drop that bomb on an American city."

"To help us prepare for, and prevent, such an attack, imagine one scenario for how this theoretically might happen," Patrick explained: Six months after the Persian Gulf attack, a terrorist atomic bomb turns lower Manhattan into a radioactive hole where Wall Street had been.

The bomb's enriched uranium had been carried into the city over many months in tiny amounts, each heavily shielded in lead and other materials to avoid detection. The weapon was assembled in place, near the top of a tall building, and set off like Iranian Improvised Explosive Devices (IEDs), by a remote signal.

The EMP from the bomb's detonation destroys nearly all nearby computers and databases, from Wall Street to the ultra-high speed stock-juggling computers across the water in New Jersey. In a heartbeat America's electronic brain center of commerce, the *New York Times* and other national media, and so much more simply vanished.

The blast's shockwave topples the Statue of Liberty.

To prevent a recurrence of such high-tech terrorism, the nation rushes to adopt heightened control and surveillance over all Americans as liberty falls across America.

With approximately 23 percent of America's 6.6 million Jews living in or near Ground Zero, this nuclear blast instantly creates a new Holocaust, a new Shoah. By targeting New York City, terrorists have struck the heart of Western capitalism and the heart of America's largest Jewish community.

Clicking-hot with radioactivity, New York City is evacuated and becomes almost a ghost town.

"It's Not My Fault"

"It's not my fault that New York City is dead," the President tells a private Hollywood party of his major donors. "My predecessors sowed these seeds, not me."

The President's words and image are secretly recorded, but the remaining Progressive mainstream media refuse to air this on their newscasts or via the Internet.

The Internet had been conceived as a communication system so interconnected that an attacker could not shut it down. Progressives used the terrorist emergency to give the government an "on-off" switch for the Web, giving the President the kind of power that dictators in other lands use to silence critics.

With the world economy in ruins, and our society now huddled inside new castles to be safe from terrorism, our country has undergone a "fundamental transformation."

The road to serfdom had delivered us to its destination.

We had been warned of the risk that the Great Debasement, with its immense debt, deficits and weakened dollar posed to America's national defense.

Condemned to Repeat It

"Progress, far from consisting in change, depends on retentiveness," wrote philosopher George Santayana. "When...experience is not retained...infancy is perpetual. Those who cannot remember the past are condemned to repeat it."

"We need to remember what happened 2,400 years ago to ancient Athens," said Professor Pat. "It was the greatest naval power in the Mediterranean Sea. This city-state spent wildly, then used force to tax its allies in the Delian League. And then came Syracuse..."

"You mean Syracuse, New York?" said Ryan.

"No, Syracuse in Greek-colonized Sicily, another democratic city-state. Athens was at war with Sparta and decided that conquering Syracuse would help it win this war," said Patrick.

"Athens felt invincible. It sent a fleet, expecting to defeat Syracuse easily."

"Instead the Athenian fleet was trapped in the harbor at Syracuse."

"Unable to fight in that narrow body of water, 200 of Athens' warships and 40,000 of its best sailors and soldiers were lost."

Athens' crushing defeat at Syracuse destroyed the myth of Athenian invincibility, the mystique of its military might, and a large part of its wealth invested in the most high-tech weapons of the time, Patrick explained.

As word of this defeat spread, allies and potential allies began inching away from Athens as they felt the tide of history and power turning in a new direction.

In a Syracuse harbor too narrow for its fighting ships, Athens fell from the zenith of its greatness – and, as recent Greek riots in Athens over debt and the Euro remind us, the greatness of Athens' Golden Age has never returned.

The Lessons of Syracuse

We remember ancient Athens as the birthplace of democracy. We need to learn and remember, as well, what went wrong in Athens that undermined its success. It has much to teach us.

Ancient Athenian democracy could be petty and partisan. Competent admirals and generals could be demoted and punished, or see slick politicians put in their place, for political reasons. This made Athenian commanders timid and indecisive in battle.

Early Athenian democracy also had its own Great Debasement of its money, replacing coins of high value with coins of cheap copper.

The famed Greek comic playwright of that time, Aristophanes, reflected on this in his 405 B.C. play "The Frogs," comparing how Athens had rejected its superior leaders and its superior money:

"We treat our best men
The way we treat our mint.
The silver and the golden
We were proud to invent,
These unalloyed
Genuine coins, no less,
Ringing true and tested
Both abroad and [in] Greece
And now they're not employed
As if we were disgusted
And want to use instead
These shoddy coppers minted
Only yesterday...."

"In 'The Frogs,'" wrote free enterprise professor Clifford Thies at Shenandoah University in Virginia in 2009, "two citizens of Athens descend into Hades for the purpose of resurrecting two well-respected politicians of the past to save the city-state from its current, corrupt rulers. The rulers are said to be like the base-metal coins in circulation, while the rulers of the past are like the full-bodied, precious metal coins that formerly circulated."

Athenian Debasement

The old gold and silver coins that built Athens into a great power "circulated abroad as well as at home because they had intrinsic value," wrote Thies.

"In contrast, the debased coins were impossible to distinguish from any counterfeit, since they had no distinctive qualities, and were repugnant to foreigners and anyone else not compelled by law to accept them, just like current politicians."

Its wars, wrote Thies, "depleted Athens' treasury. Even the gold and silver objects at the temple were melted and recast as money. Then the city resorted to debasement and to legal tender laws compelling acceptance of the debased coins at the values of precious-metal coins. Soon, the coins were recast only with base metal."

Aristophanes had grown up on the island of Aegina 17 miles from Athens, where the old values and virtues that made Greece great had not been debased by the "Progressive" values of politicized democratic Athens.

He showed that Athens' Great Debasements of its best leaders and its money were destroying its future.

Spartan Debasement

A year after Aristophanes won the equivalent of the Academy Award for Best Stageplay for "The Frogs," Sparta conquered Athens and won the Peloponnesian Wars.

Victorious Sparta was a collectivist city-state lacking Athenian democracy, science, philosophy or arts.

The one thing Spartans did superbly was fight. King Leonidas' 300 Spartans for days held at bay a Persian army of more than 50,000 at the "hot gates" called Thermopylae and became an immortal symbol of Greek independence and courage.

When fighting Athens, however, Sparta in exchange for Persian gold agreed to diplomatically recognize Persia's right to rule the Greek cities of Asia minor.

Sparta soon weakened, suffered its own defeats, and slid into the dustbin of history after its own moral and mercenary Great Debasement.

A Financial Pearl Harbor

Economic warfare began long before Sparta tactically cut off Athens' silver supply, yet today it has taken on amazing dimensions.

As Progressives have made America's economy ever-more centralized and politicized via government control, they have made us ever-more vulnerable to cyber attacks on centralized banking, corporate and government computer systems and networks.

We would be far safer under decentralized, free market supply and demand and decentralized power at the individual and local level.

As Progressives have debased our dollar and weakened our economy, they have made the United States ever-more susceptible to economic collapse.

We would be far safer putting wealth and power in the hands of the people – individuals, private enterprises and local communities.

As Progressives have fought to impose their radical notions of environmental and energy controls, they have made America ever-more dependent on foreign sources of oil that must be militarily defended. This increases the global risk of war.

We would be far safer if Progressives took their "boot" off the necks of free enterprise and let us establish energy independence based on developing our own many sources. Yet Progressives have done everything in their power to prevent us from developing and using our own hydrocarbons.

This Progressive Energy Authoritarianism is all the stranger because natural gas is one of the cleanest-burning, most environmentally-friendly fuels.

Above all, Progressives are so determined to continue using the Great Debasement as a base of their economic control that they refuse to slow government spending (except on defense) to reduce debt, even though, as noted earlier, the former head of the Joint Chiefs of Staff, Admiral Mike Mullen, describes America's debt as "the single biggest threat to our national security..."

Instead, Progressives use every crisis – including the ones they create or cause – as an opportunity to expand government, increase control over the private sector, and raise taxes. They do not let such crises go to waste.

The fragile U.S. Dollar has been engulfed, held captive and bled by every crisis around the world, and especially in the Persian Gulf. Will this be where the dollar, and soon thereafter America's military might, ends?

Nunn's Tale

"The debt problem, the deficit problem, the fact that we're borrowing so much from abroad and the fact that we are already operating in a fragile global economy, all those things combined mean that our national security is very much affected," former U.S. Senator Sam Nunn (D.-Georgia), widely respected as an expert on defense issues, in September 2012 told PBS's *Nightly Business Report*.

"The Obama budget cuts a great deal out of defense over the next 10 years...," Nunn said. "So the cliff we're facing in January [2013] is a very precipitous cut that would...really do severe damage."

"The defense budget...is not the main problem we face in the overall Federal budget," Nunn concluded. "It's entitlement programs that are simply unsustainable."

Econo-Warfare

Can economic warfare take us by surprise? It already has.

According to Defense Department consultant Kevin D. Freeman, America in September 2008 suffered a financial "Pearl Harbor," an internationally-launched coordinated computer raid designed to drain overnight trillions of dollars from our most important financial institutions, as we discussed in *The Inflation Deception.*

The resulting panic pressured federal lawmakers and Federal Reserve Chair Ben Bernanke into approving vast emergency bailout funds for banks, brokers and key corporations.

This remarkably-timed attack and resulting economic confusion persuaded voters to renew the Progressive Democratic control of Congress and to elect an unknown-and-untested anti-capitalist radical community organizer as president.

According to Freeman, this attack was the culmination of three coordinated assaults that began in 2007 with "a speculative run-up in oil prices that generated as much as $2 Trillion of excess wealth for oil-producing nations, filling the coffers of Sovereign Wealth Funds, especially those that follow Shariah Compliant Finance.

Oil prices soaring to $147 per barrel devastated the American economy, already vulnerable from the housing boom hitting a ceiling and the negative economic influence of a new and imperious Democratic Congress.

"The rapid run-up in oil prices," wrote Freeman in his 2009 analytic study *Economic Warfare: Risks and Responses: Analysis of Twenty-First Century Risks in Light of the Recent Market Collapse*, "made the value of OPEC oil in the ground roughly $137 Trillion (based on $125/barrel oil) virtually equal to the value of all other world financial assets, including every share of stock, every bond, every private company, all government and corporate debt, and the entire world's bank deposits."

Econo-Terrorism?

"This means that the proven OPEC reserves," wrote Freeman, "were valued at almost three times the total market capitalization of every company on the planet traded in all 27 global stock markets."

The second phase of this assault on U.S. financial entities began in 2008 with a series of "bear raids" against companies such as Bear Stearns and Lehman Brothers and seemed clearly coordinated to collapse the companies.

"The bear raids were perpetrated by naked short selling and manipulation of credit default swaps, both of which were virtually unregulated," wrote Freeman. "The short selling was actually enhanced by recent regulatory changes...."

"The source of the bear raids has not been traceable to date due to serious transparency gaps for hedge funds, trading pools, sponsored access, and sovereign wealth funds," wrote Freeman.

"What can be demonstrated, however, is that two relatively small broker dealers emerged virtually overnight," he wrote, "to trade 'trillions of dollars worth of U.S. blue chip companies. They are the number one traders in all financial companies that collapsed or are now financially supported by the U.S. Government. Trading by the firms has grown exponentially while the markets have lost trillions of dollars in value.'"

Jihadollars

The risk of phase three of this attack could involve "a potential direct economic attack on the U.S. Treasury and U.S. Dollar....," wrote Freeman.

"A focused effort to collapse the dollar by dumping Treasury bonds has grave implications including the possibility of a downgrading of U.S. debt forcing rapidly rising interest rates and a collapse of the American economy," Freeman wrote, noting that authorities had recently seized counterfeit U.S bonds with "a face value of $134 billion."

A year before Freeman's study, the American Center for Democracy published a detailed analysis, "The Fifth Generation Warfare," of what its analysts Dr. Rachel Ehrenfeld and Alyssa A. Lappen identified as the risk, tactics, strengths and weaknesses of what they called "Financial Jihad."

As evidenced by Freeman's study for the Department of Defense, the Pentagon is acutely concerned about the prospects for economic warfare against the United States.

The Pentagon has also been "war gaming" possible scenarios of social breakdown caused by an economic collapse. One of these war games was reportedly called "Unified Quest 2011."

Later we shall explore how the United States could protect itself from, or even win, a global economic war.

Chapter Nine
The Dollar and Its Enemies

Tipping Point # 4
The Dollar is Overthrown
As the World's Reserve Currency

"The purpose...is not to push the dollar down. This should not be regarded as some sort of...currency war."

– Janet Yellen
Federal Reserve Vice Chair
November 16, 2010

After World War II, when the United States was the last great nation whose factories were untouched by bullets and bombs, we established the ruling international monetary system known as Bretton Woods.

By this agreement, the United States pegged its dollar to gold, and many other nations pegged their currencies to the U.S. Dollar, creating a gold-exchange standard that could be used only by certain governments and central banks.

America's "Exorbitant Privilege"

The U.S. Dollar thus became the "World Reserve Currency," which meant that the most important international transactions for key goods and commodities such as oil have, from then until today, been done using American dollars as their standard unit of account and medium of exchange.

If Japan bought 1,000 barrels of oil from Saudi Arabia, the transaction had been made using not their currencies, but U.S. Dollars.

This has been a sweetheart situation for the United States because we control the making and supply of the world's currency.

Envious French President Charles DeGaulle called this "America's exorbitant privilege."

Because other nations needed dollars to engage in world trade, they had to obtain U.S. Dollars directly or indirectly from international trade by offering us things at a price we were willing to pay. This put the U.S., its enterprises, and its dollar in a strong bargaining position as buyers.

"Fully 85 percent of foreign-exchange transactions world-wide are trades of other currencies for dollars," wrote University of California Berkeley Economics Professor Barry Eichengreen in 2011.

"What's more, what is true of foreign-exchange transactions is true of other international business," Eichengreen continued. "The Organization of Petroleum Exporting Countries [OPEC] sets the price of oil in dollars. The dollar is the currency of denomination in half of all international debt securities. More than 60% of the foreign reserves of central banks and governments are in dollars. The greenback, in other words, is not just America's currency. It's the world's."

We were also in a strong position as sellers. The U.S. did not need to buy secondhand dollars from middlemen before doing international business. We were dealing with the entire world in our own national currency, which (except for its redeemability in gold by certain foreign central banks) our government could print in whatever quantities it wished.

"But astonishing as that is," wrote Professor Eichengreen in the *Wall Street Journal*, "what may be even more astonishing is this: The dollar's reign is coming to an end."

Eichengreen predicts that the dollar's value will "fall by roughly 20 percent" when it ceases to be the world's sole Reserve Currency. Americans will have to pay significantly more for imported goods, and will see their standard of living decline.

"If the commodity-rich enemies of the U.S. want to cause problems in this country, they don't need to fire a single shot," wrote Brian Rogers, now an Adjunct Professor at New York University's Stern School of Business, in 2011. "All they have to do is start selling their products priced in gold. The end of the reign of the USD [U.S. Dollar] as a reserve currency will follow quickly."

Nixon Shock

On August 15, 1971, President Richard Nixon shocked the world by unilaterally ending the U.S. Dollar's convertibility to gold.

Both Mr. Nixon and his predecessor President Lyndon Johnson had printed vast amounts of paper money to bankroll a costly war in Vietnam.

At the fixed convertibility of the Bretton Woods agreement, each U.S. Dollar could in 1971 be exchanged for $1.34 worth of gold at its free market price.

Fearful that the French central bank was about to trade a mountain of paper dollars for all the gold in Fort Knox, President Nixon slammed shut America's gold window for such conversions.

Spending Unchained

Adrift and anchorless, the dollar and other currencies that had been pegged to it were floating. Each would find its own market value.

The floating U.S. Dollar quickly lost a third of its value, and by 1980 had lost half its 1971 purchasing power. A tidal wave of inflation was devouring the American economy, and under President Jimmy Carter interest rates soared into double digits.

One of the biggest reasons for this plunge in the dollar's value was that even the limited gold standard of Bretton Woods had put golden manacles on politicians that constrained their spending.

Thomas Jefferson spoke for other Founders when he said of politicians that we must "bind them down with the chains of the Constitution."

One of the strongest of those Constitutional chains was that American money be based on precious metal. The Founders knew what happens when reckless lawmakers can simply print all the paper fiat money they desire.

President Nixon unchained the government from monetary constraints. Lawmakers were now free to spend, and freely did so. Entitlement spending exploded.

As a result, what one pre-Nixon Shock dollar could buy would by 2011 cost $5.65.

The Dollar's Nine Lives

"I told you that the U.S. Dollar is like a cat. It's had nine lives and is about to lose its last one," said Professor Pat.

"So what are they?" asked Ryan.

"The first was paper fiat money, the Continental that our Founders had to use during the Revolutionary War because they were cut off from British money," said Patrick.

"As we discussed, the second was the Bank of North America dollar that replaced the Continental. The third were the Constitutional dollar of gold or silver, along with those issued by the First and Second Banks of the United States."

"America's fourth dollar was the fiat paper Greenback issued during the War Between the States," said Patrick. "The fifth was the gold standard dollar from 1879 until the Progressives began replacing it with Federal Reserve money. The Sixth dollar is the fiat paper Federal Reserve Note that, in an even weaker version, is still with us."

"Incidentally," said Patrick, "Abraham Lincoln proposed having the U.S.

Government issue its own currency directly to avoid having to pay interest to anyone, as American taxpayers have paid to the Federal Reserve since 1913. The Fed, of course, says that it rebates all such nominal annual 'profits' – more than $80 Billion a year in recent years -- back to the Federal Government. The Fed, however, has opposed efforts to subject it to an independent audit and has shadowy relationships with banks and other financial institutions around the world. Who knows how much its activities generate?"

"Under President John F. Kennedy the government issued a limited quantity of United States notes that were not necessarily Federal Reserve Notes. They came from the government itself and were printed to ease the transition away from silver certificates."

"The last dollars that American citizens could redeem for gold died in 1933 and for silver in the mid-1960s. I count them together as the Seventh U.S. Dollar, the precious metal dollars," said Patrick.

"After World War II the U.S. created Number Eight, the Bretton Woods dollar, which certain foreign central bankers were permitted to redeem for gold. By pegging their currencies to the U.S. Dollar, many other nations adopted a last vestige of the gold-exchange standard. This is what established the U.S. Dollar as the current world Reserve Currency."

"This is what President Richard Nixon killed in August 1971, when he slammed shut the gold window of the Bretton Woods dollar. This left us with an anchorless, adrift Dollar Number Nine, the floating paper fiat dollar that a former member of the Federal Reserve's Board of Governors John Exter nicknamed the "I.O.U. Nothing" dollar. When it dies, the dollar will be no more – unless we can bring an older, better dollar back to life," said Patrick.

Rivals Arise

If the U.S. Dollar is no longer anchored in gold, and if the U.S. Government and Fed debase its value by printing dollars in irresponsible quantities, then why should the dollar remain the world's Reserve Currency?

As a debased and sinking Reserve Currency, the dollar has increased economic uncertainty, instability and insecurity in global markets.

Yet thus far, no rival currency has been strong enough to challenge and take the dollar's place. The major competitors are themselves also fiat currencies of uncertain strength and reliability.

One is the Euro, launched in 1999 by 15 of the 17 current European nations as their new common currency. Of the 27 nations in the European Union, all except the United Kingdom and Denmark are legally bound at some future point to join the Eurozone and adopt this currency.

The entire European Union has less than half the land area of the United States, yet in 2011 it had a combined population of 504 million, a labor force of 228 million, and combined Gross Domestic Product (GDP) of $15.65 Trillion.

The United States, by comparison, in 2011 had a population of only 314 million, a labor force of 154 million, and a GDP of $15.29 Trillion.

Euro Nations

Why is the Euro not an obvious contender likely to replace the weakening U.S. Dollar as the world's Reserve Currency?

And why was the European Union's 2011 per capita GDP almost $15,000 lower than America's $49,000?

A zealous American might say that the best, most enterprising and adventurous Europeans moved during the past 400 years to the United States. Americans embody the cream of Europe's gene pool.

As we discussed in *The Inflation Deception*, part of American exceptionalism may, indeed, come from a higher proportion of a specific restless "entrepreneurial" gene that may exist disproportionately in American DNA.

The 20th Century's two World Wars also took a heavy toll of Europe's best, brightest and most patriotic, both on the battlefield and as a prod to emigration.

American Welfare

After World War II, the United States did several things that have altered

European politics, economics and culture.

In postwar Europe President Harry Truman authorized the Marshall Plan, an outpouring of food, medicine and reconstruction aid worth in today's inflated currency perhaps a trillion dollars.

Humane and well-intended, this aid for some beneficiaries became an addictive dose of government dependency.

During postwar reconstruction, the United States established forward U.S. military bases in Western Europe and Japan, and pledged that our nuclear arsenal would protect these nations free of charge from aggression by the Soviet Union.

This pledge that an attack on one was an attack on all became the cornerstone of the Atlantic Alliance and its force structure NATO, the North Atlantic Treaty Organization.

Uncle Sam's Umbrella

America's nuclear umbrella protected a shattered Western Europe, but it also transformed politics there.

Nations that previously paid for their own defense now left this heavy responsibility mostly to the United States and the half-million U.S. troops stationed in Europe.

This Pax Americana gave Western Europe the longest period free from war that it had known since the Roman Emperor Augustus 20 centuries earlier.

What these nations saved on defense, they mostly spent on creating socialist welfare states.

With the Soviet Red Army occupying Eastern Europe, moderate American Progressives saw socialism and welfare as ways to inoculate Western Europe against Communism.

Many American conservatives perceived this differently. Comic writer P.J. O'Rourke crystallized their concern. "Liberalism," he later wrote, "is just Communism sold by the drink."

American Progressives soon developed Euro-envy and an endlessly-repeated chant: "Why can't we have socialized medicine and welfare benefits like France or England?"

We could not, of course, because during the five decades of the Cold War against the Soviets, we chose to be the grown-ups who paid for Western Europe's and Japan's national security in a dangerous world.

Western Europeans enjoyed the luxury and ease of having American taxpayers pay for what otherwise would have been their biggest national expense.

America the Enabler

By this act of generosity, however, we created in Western Europe a huge constituency of public employees and welfare recipients dependent on government money.

The democratic votes of such Europeans empowered socialist Social Democratic, Labourite, Green and Communist political parties that were sometimes quick to bite the American hand that fed them.

In war-weary Western Europe, people living under our nuclear umbrella and the nearby Soviet Union's red shadow indulged themselves for decades as their culture and faith crumbled and their fertility rates declined to below-replacement levels.

In 2010 the Total Fertility Rate across the European Union averaged 1.58 children born per woman. Just to sustain a nation's population requires 2.1 children per couple.

In demographic terms, the old European peoples who gave the world Greek democracy, Roman Imperial law, English government and the Scottish Enlightenment will soon be no more.

The average Muslim couple in Western Europe today has 3.84 children. Over time Europe will become Eurabia as Muslims conquer this continent by making love, not war.

In the United States, by comparison, the Total Fertility Rate in 2012 is estimated by demographers to be between 1.87 and 2.06, with a disproportionate increment of this "reproductivity" coming from Christian Latino immigrants. About half the children age one or younger in the U.S. in 2012 are minorities. The U.S. will survive the mysterious lack of fertility that is dooming advanced nations from Japan to Russia to Western Europe.

European Christian churches now stand nearly empty, while Muslims from what once were European colonies have become sizable and influential voting minorities in Germany, France, the Netherlands, Great Britain and several other European nations.

Europe, by American standards, has become a place of high taxes, soaring gasoline prices, chronic high unemployment, and a still-struggling attempt to create a united states of Europe and a rival to the U.S. Dollar's status as the world's Reserve Currency.

Germoney

The Euro may have begun as a disguise for the Deutschmark. It may be Germany's third attempt launched in the 20th Century to end the 400-year-old civil war in Europe by dominating the continent economically.

Germany depends on exports to thrive. The Euro, and the loans made by German banks to nations using it, turned the other 16 Eurozone nations into prime markets for Germany's goods.

Yet Germany is a surprisingly small locomotive to pull Europe's entire diverse economy. And it is in demographic decline, with a Total Fertility Rate of only 1.41.

Even after the reuniting of the East and West halves it was split into after World War II, Germany is a country smaller than the U.S. state of Montana.

A strong 33 percent of Germany's land is arable, however, and rivers gave it a kind of highway transport system for boating goods since ancient times.

Little Giant

Germany as of 2012 had 81 million people, a labor force of 43.6 million, and a GDP in 2011 of $3.1 Trillion – roughly one-fifth of America's.

Neither six decades as a Social Democratic welfare state in West Germany, nor four decades under Soviet rule in East Germany, could destroy Germany's remarkable work ethic. Its per capita GDP in 2011 topped $38,000.

While President Barack Obama chose to fight the Great Recession with massive Keynesian stimulus in an effort to tax, spend and borrow America back to prosperity, Germany instead turned from left to right.

Germany moved back towards capitalism, somewhat reducing the size of government and the welfare state, and strengthening its economy through savings and austerity.

This rightward path has produced stronger economic growth in Germany than in the United States.

Nightmares of Weimar Hyperinflation

Germany understood firsthand the inflation nightmare of its Weimar Republic era after World War I and how the values-destroying hyperinflation paved the way to Adolf Hitler. We discussed this in extensive historical detail in our book *Crashing the Dollar*.

German Chancellor Angela Merkel implored President Obama not to take the Weimar path of deliberately inflating the currency, but to join Europeans in a program of austerity.

(In much of Western Europe, of course, "austerity" does not mean reducing the obese size of government; it means raising taxes as high as necessary to pay for such government.)

Mr. Obama rebuffed this plea from both Germany and the United Kingdom. While Germany moved towards capitalism, Mr. Obama turned the nominally-capitalist United States towards failed Eurosocialism.

Gimme PIIGS

What hectors Germany is that the Eurozone includes nations whose work ethic falls far short of Germany's.

These profligate nations – the so-called "PIIGS": Portugal, Italy, Ireland, Greece and Spain – used their expanded credit they were offered after joining the Eurozone to go on wild spending sprees and take on fantastic debt loads.

Having run up stratospheric bar tabs, these Gimme PIIGS now want to stick someone else – preferably the taxpayers of Germany or the United States – with the bill. These liberal-spending nations have little inclination to tighten their belts if they can manipulate somebody else into paying their debts.

For German workers and taxpayers, such attempts to stick them with the bill seem highly unethical. In effect, the PIIGS are holding the survival of the Euro, which Germany needs, as a hostage for ransom.

This is producing drag on the German economy and on the survival prospects of the Eurozone and even of the Euro itself.

TransAtlantic QE

In mid-September 2012, the European Central Bank (ECB) announced its own Quantitative Easing to push back rising unemployment and stimulate the weakening European economic situation.

Scarcely hours later, American Federal Reserve Chair Ben Bernanke launched QE3, offering an open-ended $40 Billion per month injection of conjured Fed money into the economy in general and the housing market in particular.

QE3 was added on top of the Fed's ongoing Operation Twist, already a sort of QE3 based on swapping long-term and short-term debts to keep interest rates down that was scheduled to end on December 31, 2012. During the last months of 2012, QE3 and Operation Twist combined could print $85 Billion per month out of thin air and add it to the economy.

This huge coordinated transatlantic stimulus inevitably will generate a tsunami of high inflation when it finally reaches the marketplace that could inundate Europe and the U.S. The only uncertainty is when. Any crack in the economic castles of both continents will unleash it and potentially wash away both the U.S. Dollar and the Euro.

The FrankenEuro

Some experts are not waiting for the Euro to die. Their reason: it is already dead.

Forbes Magazine columnist and academic economist Mark Hendrickson argued in July 2012 that the Euro was never a natural, organic living currency. It is, he writes, "a Frankenstein currency" that was "grafted together from pieces of 17 (so far) dying fiat currency trees of different species... stitched together from pieces of the deutsche mark, French franc and Italian lira fiat currencies."

Those now using the Euro currency "are trapped," he writes.

"Either member countries will abandon the euro, in which case banks, governments, businesses and individuals go through a wrenching period of defaults, write-downs, 'haircuts' and bankruptcies, or they lurch onward toward an unviable fiscal union in which Germany, Finland and the few relatively solvent economies are crushed under the unsupportable weight of being expected to bail out the relatively bankrupt countries."

The Zombie Euro

Europe is now haunted by the Euro Zombie, wrote *Wall Street Journal* MarketWatch columnist Matthew Lynn in August, 2012.

Real money, he wrote, is about "being a universally accepted medium of exchange, a store of value over time, and a way of facilitating trade over long distances. That was why money evolved. And the euro doesn't really meet those criteria any more."

Interest rates vary wildly across the supposedly-united economic Eurozone, Lynn writes. Countries fear using Euros vis-a-vis Greece out of fear that its old Drachma may suddenly replace Euros. Wealth in the Eurozone is fleeing to Switzerland and London.

"The euro still has the notes and coins," Lynn writes. "But in almost every other sense, it is no longer really a currency."

"When a currency stops working the damage done to the economy is immediate. Trade stops flowing. Investment gets postponed. Capital flees. Very quickly, unemployment starts to rise, and output declines," writes Lynn.

"That is precisely what is happening in the euro zone right now."

Hard-pressed countries such as Spain and Italy will return to their pre-Euro currencies, Lynn predicts, and their economies will rapidly improve once they are freed from the yoke of a Euro currency they use but whose monetary policy is out of their control.

Serfs of the New Sparta

Such countries that remain yoked to fiscally-disciplined northern European nations such as Germany will be like serfs, the Helot slaves of today's Sparta, almost as if they lived in German colonies within Europe itself.

This is why the Frankenstein / Zombie Euro is unlikely any time soon to replace the U.S. Dollar as the world's Reserve Currency....unless the dollar itself continues down the path to becoming a dead, ghostly Zombie currency itself.

At least two other currencies are spoken of as plausible Reserve Currency replacements for the weakening U.S. Dollar. One is a new global currency associated with the International Monetary Fund (IMF), an entity created by the Bretton Woods agreement, or with the United Nations.

The other possibility is the currency of the world's #2 economy, China's Yuan.

Today's Euro seems to need its own regular injections of stimulus money to stay on its feet. It may now be an even less stable global Reserve Currency than the dollar. Because of the heavy trade links between the Eurozone, China and the U.S., a collapse of the dollar, the Euro or the Yuan/ Reminbi could set in motion a domino effect that could severely weaken the other two. During 2011 and 2012 the Euro fell as much as 30 percent relative to the Yuan and U.S. Dollar.

The UN-Money

Ever since John Maynard Keynes proposed what he called the "Bancor," Progressives have fantasized over a global fiat currency churned out by a global government such as the United Nations.

The question is: who would run the central bank controlling such a global currency? Would it be in the hands of capitalist countries, socialist satraps, or communist nations such as Cuba, just to make it "fair?"

What mix of ideology, personal greed and national/ethnic/religious blocs would control such a global currency in a United Nations where the 84,000 people of the Seychelles islands in the Indian Ocean have as many votes in the U.N. General Assembly as the United States?

It is difficult to imagine the world rushing to invest its trust or fortunes in a United Nations currency.

Go East, Young Man

With America's economy being devoured by a black hole of debt, however, the dollar will not long remain the world's Reserve Currency. It will either lose this status or, more likely, be forced to compete for this lucrative status with one or more other currencies.

China, with 1.34 billion people, has imperial ambitions to expand its power and influence, both military and monetary, first in the West Pacific region and then around the world.

China's ambitions are disquieting to its neighbors, who feel the sway of its growing might and designs on global resources, including South China Sea oil and shipping lanes.

China Doll-ars?

"Teach your children Spanish and your grandchildren Chinese." This advice is nowadays heard from some Americans. When betting on who will run the world's future, many put their chips on China.

China may no longer seem as Marxist as it once was. Confucian culture produced overseas Chinese who are among the greatest capitalists and business people of all Asia, and whose work ethic surpasses the increasingly-slacker cultures of the West.

The People's Republic, however, practices a variety of quasi-state capitalism that has a great deal of collectivist government involvement. As the economy has gone into contraction recently, loans from government banks have begun going mostly to government enterprises, cutting off most private sector companies from capital they previously counted on.

China in the fall of 2012 was preparing for a turnover of most members of the Communist Party's ruling body, a change that has happened once every 10 years in recent decades. This could produce a step backwards away from the Communist nation's economic "convergence" with the West.

China has kept the value of its currency artificially low for international trading advantages. For this, as well as recent indications that China's economy is weaker than used to be assumed, the Yuan/Reminbi is less than an ideal candidate to replace the dollar as the world's Reserve Currency.

China, however, has in recent years signed as many as 18 important bilateral trade agreements. Some of these, including a deal with Russia, have included agreement that the countries trade with each other using their own currencies, not dollars.

China is reportedly in the process of boosting this to as many as 36 bilateral deals, and such arrangements that bypass the dollar's Reserve Currency status and power will weaken the dollar.

Green and Red

"The moment China has acquired everything it needs from the United States, it will turn against you to destroy your economy," an Asian economist told one of us matter-of-factly over supper in Singapore.

On January 23, 2012, an ominously auspicious time for such earth-shaking events began as Chinese New Year throughout the world welcomed the Year of the Dragon.

In China's ancient calendar, one year in a cycle of 12 different animal years is dominated by this mythical beast, which, unlike dragons in the West, is believed to bring good fortune, prosperity, success, volatility, energetic change and power.

Dragon years typically bring five percent more births because couples think it lucky to have Dragon babies. And millions of Chinese for thousands of years have thought of themselves ethnically as the people of the dragon.

The dragon was the symbol of China's first emperor, and of all his successors who ruled half the world from the Dragon Throne of the Middle Kingdom. Each Year of the Dragon was propitious for Chinese imperial power.

Power Shift

At nearly the same time as America's 2012 presidential election, this Year of the Dragon marks a major transfer of power in China's leadership. The 18th National Congress of its Communist Party was expected to replace President Hu Jintao, Premier Wen Jiabao, and seven members of its nine-member ruling Politburo. These changes will be mirrored down the hierarchy of government from top to bottom as a new generation of leaders takes power.

China under Hu Jintao was already exercising its new power as the world's second largest economy and biggest foreign buyer of U.S. debt, more than $1.35 Trillion worth, playing enabler to the spendaholic addiction of our politicians.

The People's Republic of China has likewise become the biggest foreign buyer of Japan's debt, and late in 2011 the two nations announced an agreement to trade with one another in their own currencies.

China has been shifting away from buying American debt into buying real things such as deposits around the world of copper and "rare earths" needed to make advanced magnets and high-tech materials. It has been buying prime farmland, from Iowa to Zimbabwe. And China has become the world's biggest buyer of gold, while selling none from its own mines. Central banks around the world have also recently been buying gold at the highest rate in 40 years.

One might almost wonder if China is preparing to create its own commodity-based money system.

China out-negotiated President Obama's Administration to win two choice oil deals with Brazil's state company Petrobras. Canada's Prime Minister visited Beijing in February 2012 to discuss China's acquisition of tar sand oil in Alberta that Mr. Obama has blocked U.S. efforts to import. A $15 Billion Chinese purchase of a key Canadian owner of this oil was announced in 2012.

In January 2012 China, a nuclear weapons nation, signed an agreement to cooperate with Saudi Arabia in developing nuclear technology for "peaceful purposes." Such things happen when a vacillating American president reinforces the growing global impression that China is a superpower on the rise...and America's power is on the wane.

China recently became the biggest trading partner of Australia, and has discussed helping with the financial morass of Eurozone nations, who are customers for roughly 40 percent of China's exports.

Sinking West, Rising China?

How must it look from Beijing to see Western nations that once occupied China as a colony now begging for bailouts from China?

"If you look at the troubles which happened in European countries," Jin Liqun, supervising Chairman of China's sovereign wealth fund the China Investment Corporation, told a reporter, "this is purely because of the accumulated troubles of the worn-out welfare society.... The labor laws induce sloth, indolence, rather than hardworking.... The incentive system is totally out of whack.... A welfare society should not induce people not to work hard."

In opening American relations with Communist China, anti-Communist President Richard Nixon gambled that we could turn China capitalist before making it wealthy and powerful enough to destroy us.

Following the 1989 massacre of students at Tiananmen Square in Beijing, China's "paramount leader" Deng Xiaoping in 1992 called for opening the economy. He described China's "leftist elements" as more dangerous than its "rightist" ones. He unleashed thousands of entrepreneurs by proclaiming that "To get rich is glorious."

A Changing Dragon

Until recently, it appeared that the U.S. might win President Nixon's life-or-death wager.

In January 2012, however, President Hu Jintao declared that "hostile forces abroad" are trying to "Westernize" and "divide" China with their cultural influences. He ordered, among other things, a crackdown on television shows that were "excessively entertaining," reflecting Western values and morals.

Analysts suggest that Hu's actions are intended to deflect growing popular anger in China at low wages, inflation, economic growth that China says recently slowed to only 8.9 percent per year, and government corruption and land expropriation by Communist officials.

State Capitalism

The high tide of entrepreneurial capitalism permitted by the Communist government might now be ebbing. The private sector employs 80 percent of China's workers, but it is now shrinking because capitalists receive only 20 percent of loans from the nation's government-controlled banks.

Fully 80 percent of bank loans now go to unproductive government enterprises whose state planners have built industrial ghost towns and other under-utilized infrastructure. This is bleeding the capital out of Chinese capitalism.

The world's largest military force, the Chinese People's Liberation Army, also gets ample funding to develop cyber warfare to attack Western computers and laser cannons to disable the spy and positioning satellites that guide America's cruise missiles and other weapons.

China has bought and put to sea its first aircraft carrier, positioned more than 1,200 intermediate range missiles within striking distance of Taiwan, and begun claiming ownership of islands in the China Sea as if the Asian Pacific is its hegemony.

"The borrower is slave to the lender," the wisdom of Proverbs 22:7 tells us. Is a United States dependent on China's money willing – or able – to stop Communist China from invading our capitalist ally Taiwan?

National defense and America's ability to fight two simultaneous wars seem to be the only areas of spending our politicians seem willing to cut, even as President Obama recently announced a shift in America's military focus from Europe and the Middle East to the Asian-Pacific Region.

The Art of Economic War

"There are two ways to conquer and enslave a nation," warned America's second President, John Adams. "One is by the sword, the other is by debt."

Ancient China's military strategist Sun Tzu's *The Art of War* teaches that wars are won by using *everything* as a weapon, from psychology to a rival's economic weakness and debt.

China has announced that its imperial ambition extends beyond the Pacific Ocean and planet Earth. It has already begun plans to land Sinonauts on the Moon around 2020 and on Mars soon afterwards.

China's push to lead the world in scientific research and development, is being funded by American consumers – and by taxpayers who pay interest on the money China lent us that our politicians are giving back to China as free foreign aid. Money being fungible, one could say that our interest payments now fund much of the Chinese military.

The U.S. and China may continue to converge economically into what Harvard University economic historian Niall Ferguson has called "Chimerica." Seen in imperial terms, the United States allows General Electric and other companies to establish high technology facilities in China, yet we also have become a major supplier to China of raw materials – the traditional position of a colony.

We need to appreciate China's understandable upset that we might repay our debts to them with inflation-devalued dollars worth much less than the money they lent to us. Were we in their position, we would have similar feelings.

China's Fear

We also need to understand the cable (discussed in the September 14, 2011, *Zerohedge.com*) that President Obama's then-Ambassador to China Jon Huntsman sent in February 2010 explaining a great fear of Beijing's rulers – that the U.S. could suddenly re-value our dollar, then peg its new value to gold.

Gold could be America's winning weapon in the emerging new "Gold War" against China's plan to dominate the world economy.

China has remained a powerful culture and civilization for thousands of years, largely because of Confucian conservatism, family values, thrift and work ethic. China also has the patience of an ancient culture that thinks in terms of dynasties and centuries. It does not have a short-sighted Western focus on maximizing profits next Quarter, or winning the next election, by doing things that will make things worse in the long run.

One manifestation of this has been the ancient Chinese habit of giving and storing gold to protect families and the economy against bad times.

In the entire year 2010, China bought 120 metric tons of gold. By November 2011, its gold purchases had jumped more than six-fold to 103 tons *per month*.

China in 2012 surpassed India, which for cultural reasons historically had been the world's biggest gold consumer.

What have the Chinese seen coming in the Year of the Dragon and beyond that has them rushing to buy gold? What do they know that we do not?

Do they expect to rule the world after the collapse of Western economies built on intrinsically worthless paper fiat money and inflation? Their history here goes back farther than anyone else's, because China invented both paper and paper money.

Money Turns to Paper

While much of Europe faced the Dark Ages, China's Song Dynasty invented paper and, by the 7th Century A.D., what we think of as paper money.

Using wood block printing, the Song at first produced what amounted to promissory notes that could be redeemed for metallic coins. When coin copper ran short, the government began issuing paper that simply took the place of hard money.

In some ways this paper currency was remarkably modern. By 1107 A.D., it was printed on a special paper with intricate designs and six different colors of ink to make counterfeiting difficult.

By 1175 A.D. China's government was operating at least four factories in different cities to churn out paper money. One of these factories, records show, employed 1,000 workers.

In other ways, the early Song notes seem odd to us today. Distinctly different paper notes were printed for use in specific regions of China. Each note carried a time limit of three years, which meant that they could not be saved, hoarded or hidden from the tax collector for a longer period than that while retaining their store of value.

Sometime around 1268 A.D. the Southern Song Dynasty began printing a single China-wide currency convertible to gold or silver, but their days were numbered.

Mongol Money

By 1279 the Song's last defenders were defeated by Mongol troops of the grandson of Genghis Khan, Kublai Khan, and his Yuan Dynasty. Kublai at first printed the restricted currency used by the Song. The Yuan soon created a flood of currency unbacked by metal, yet without time limits on when it could be spent.

The Mongol ruler required his subjects to accept his currency at face value.

"The Chinese government confiscated all gold and silver from private citizens and issued them paper money in its place," wrote anthropologist Jack Weatherford.

"Even merchants arriving from abroad had to surrender their gold, silver, gems, and pearls to the government at prices set by a council of merchant bureaucrats."

"The traders then received government-issued notes in exchange," wrote Weatherford.

"Ghost Money"

Marco Polo, a Venetian merchant who visited Mongol-ruled China, "saw clearly that this system of paper money could work only where a strong central government could enforce its will on everyone within its territory," wrote Weatherford.

Inflation began rising as succeeding rulers enjoyed the royal prerogative of manufacturing money.

The Chinese took to calling paper money "wind money" because its value could easily blow away.

"Some people in Asia burn joss paper, also called ghost money, on the Lunar New Year, to give their deceased ancestors something to spend in the afterlife," wrote Princeton University philosopher Daniel Cloud.

"Because ghost money doesn't represent a claim on any actual goods or services in *this* world, there is no reason for its issuers to exercise any particular restraint, and in Singapore it is possible to find notes issued by the First Bank of Hell, with the mythical Jade Emperor's picture on the front, in denominations ranging into the millions and billions of dollars."

"Perhaps we're counting on this charming tradition," wrote Cloud at *Zerohedge.com* in 2010, "to make Asian investors comfortable with the prospect of continuing to add to their holdings of European and American sovereign debt, despite the obvious fact that the money they've already lent us is money they'll never get a chance to spend in this life."

The People's Republic of China continues to hold more than $1.1 Trillion in U.S. paper promises, despite clear evidence that the U.S. Dollar itself is now being so hollowed out that it is turning into ghost money.

If American dollars are a key part of the reserve China uses to support its Yuan / Reminbi, then their currency, too, may be little more than ghost money and hence too unworldly to be the next world Reserve Currency.

"Anyone who [in today's fiat currency-denominated paper] holds a lot of sovereign debt," wrote Cloud, "is at risk of eventually discovering that it is fairy gold, ghost money, mere joss paper that didn't even correspond to any pile of goods and services actually available in this world."

Balance of Trade

By 1455 the Ming Dynasty, to stabilize its economy, banished paper money and curtailed most international trade.

This Dynasty also destroyed its large imperial Chinese fleet that decades before Columbus had traded Chinese goods as far away as the island of Madagascar off the coast of southeast Africa.

Former British Royal Navy submarine commander Gavin Menzies has laid out a speculative case that ships of this Chinese fleet might have reached Europe and influenced the Renaissance, and might have reached the New World more than 70 years before Columbus.

Had the Ming Dynasty retained its large fleet and sailed to the New World more than three generations ahead of Columbus, this might have changed human history in many ways.

Had they continued, you might be reading this book – or more likely a very different book from a different culture and history – right now in Chinese.

The International Monetary Fund in April 2012 predicted that China's economy will surpass that of the United States by year 2016, and that the era of America's global supremacy is about to end.

China's treasure fleet has arrived to conquer the West more than 500 years late, but many now believe that a long-postponed age of Chinese world supremacy is beginning.

Others look at declining productivity in China and wonder how much of China's claimed success is real – and how much is illusion.

The Race to Debase

The bad news for the Euro, Yuan and U.S. Dollar is that we continue to live in an upside-down reality where the central bankers and politicians scheme to drive the value of these currencies not up, but down, to make them worth less.

This makes a nation's exports cheaper and imports more expensive, both of which supposedly benefit domestic manufacturers by driving sales their way.

Yet this is a race to the bottom, a deliberate debasement of people's savings and trust, a contest to beggar our neighbors that ultimately makes all of us poorer.

Some analysts now call it "the Race to Debase."

Such are the currency wars that continue to be fought among Earth's governments and central bankers as part of the inherent manipulation of fiat money.

Those who deliberately cheat today's savers and betray future generations by conjuring easy money and false prosperity now do not deserve positions of public trust. They should not be policy makers able to shape the currencies on which people rely. No such deliberately-debased money deserves to be the world's Reserve Currency.

In global currency wars, every human being on the planet eventually loses.

The good news is that one kind of money is available that transcends short-sighted political manipulation, as we shall learn. If this money becomes the world's Reserve Currency, everybody can win.

"Inflation can be pursued
only so long as the public does not believe
it will continue. Once the people
generally realize that the inflation
will be continued on and on
and that the value of the
monetary unit will decline more and more,
then the fate of the money is sealed.
Only the belief that the inflation
will come to a stop
maintains the value of the notes.

– Ludwig von Mises
Austrian economist

Part Three
How to Survive
The Death of the Dollar

Chapter Ten
Ghost Money:
After the Dollar's Nine Lives

"Paper is poverty....
It is only the ghost of money
and not money itself."

– Thomas Jefferson

"There will be a time...
when physical money is just going
to cease to exist."

– Robert Reich
Economist and
former Secretary of Labor

A specter is haunting the world – the specter of electric money.

Thomas Jefferson saw paper fiat money as only the ghost of real money, which to him and America's other Founders was supposed to be specific weights of silver and gold.

We are already entering an age when the dollar will end the last of its nine physical lives, and die in both its metal and paper physical incarnations.

The dematerialized dollar's ghost will then haunt our economy. It will linger as an apparition, conjured by invisible electrons dancing through computer circuits at almost the speed of light.

Such disembodied, phantom money is already causing a huge revolution in our economy, our society, and our most basic ideas about life and liberty.

This revolution will be far bigger than the switch from the intrinsically-valuable precious metal money of our ancestors to today's faith-based fiat paper currency whose value is forever falling.

The Cashless Future

In today's America, only 7 percent of transactions are still done using the old-fashioned tangible, physical money. Most of these cash purchases are small, such as hamburgers, candy bars, or the liquid candy of soft drinks.

In high-tech Sweden, where the government encourages steps toward a "cashless society," only 3 percent of transactions still use coin or paper currency.

To ride the bus in Sweden's big cities requires a token or electronic payment. No cash is accepted.

Some Swedish banks reportedly keep no cash on hand at all, and accept none from depositors. They might take a check, yet most deposits and bill-paying are now done via electronic funds transfer.

Millions of America's Social Security beneficiaries have chosen to enjoy the convenience of monthly payments sent digitally directly into their bank accounts. The same, say experts, will soon be done with almost everyone's paychecks and tax payments.

Clean Capital

The cashless age of such ghost money will be cleaner and healthier, advocates promise.

You will no longer handle dirty coins or currency passed to you through the sticky fingers of anonymous drug users or sniffling flu sufferers.

The cashless age will be safer, say advocates. You need not carry cash, or eventually even credit or debit cards – the morning stars of this dawning new age.

Soon your transactions will be validated by some biological identifier that only you carry. Fujitsu has already invented a scanner that reads your open palm's blood vessels – a pattern more unique to you than your fingerprints or signature – to confirm your identity.

In this cashless future, therefore, small-time thieves will no longer be able to steal your money by picking your pocket, snatching your purse or robbing the cash register at a small merchant.

You and your virtual money will have become virtually one. It will be unspendable without your active participation and approval.

Moneyless in Eden

Serpents will emerge in this future cashless Eden, warn critics.

When a nation's entire economy exists mostly as mere manipulable impulses and traces in the cyber code of fallible computers, tech-savvy thieves will devote enormous resources to deciphering how to destroy, impair or steal vast amounts of wealth or credit, not to mention the vulnerability of our national power grid and other systems. This, as we shall see, is already happening.

In a cashless society, government will demand the power to monitor and then to reach its greedy hand into every public and private transaction, large and small.

A cashless society could therefore become a naked society to the prying eyes of bureaucrats, tax collectors and partisan political rulers.

The death of the dollar thus could also mean the death of individual privacy.

Google Bucks

Needless to say, a centralized Progressive government will want the emerging cashless society to give even more power to the government and less power to the people.

"We had various proposals to have our own currency, which we were going to call Google Bucks," the Internet giant's Executive Chairman Eric Schmidt told a 2012 industry gathering.

The company dropped this idea, said Schmidt, because of "some issues with peer-to-peer money. It turns out that it's in most cases illegal.... The reason that it's illegal is that governments don't trust it because of the issues of money laundering and so forth, *and the central banks want to control it*." (emphasis added)

Google retreated to a simpler joint project with the telephone company Sprint called Google Wallet. In 2012 a similar endeavor, Isis, by cellular companies Verizon, T Mobile and AT&T was expected to launch.

Both Isis and Google Wallet are designed to facilitate product marketing and targeting, and easy consumer buying, via cell phones. Critics say that thus far these approaches seem to be little more than the old plastic credit card in the shape of a cell phone, yet both might quickly evolve into lucrative enterprises.

Off the Books

Some of the government's concern about digital money is understandable.

A 2011 study by Edgar Feige of the University of Wisconsin-Madison and Richard Cebula of Jacksonville University in Florida concluded that between 18 and 19 percent of total reportable income in the United States is effectively off the books, hidden from the government.

This income – from drugs, prostitution, private gambling, home repairs, and a thousand other things paid in cash – could, Feige and Cebula estimate, have harvested half a trillion dollars in tax revenue for the government.

Cash makes it easy for criminals to thrive in an economic underworld of untraceable illicit transactions. A cashless society, say advocates, would drag this underworld out of the shadows and into disinfecting sunlight. In a cashless society, they say, crime would not pay as well as it does today.

Civil libertarians, however, are troubled that cashless crusaders now seem to be demonizing cash.

The Stigma of Cash

"We're trying to use industrial age money to support commerce in a post-industrial age. It just doesn't work," David Birch, a director of Consult Hyperion, a firm that specializes in electronic payments, told *Slate* Magazine.

"Sooner or later," Birch continued, "the tectonic plates shift and then, very quickly, you'll find yourself in this new environment where if you ask somebody to pay you in cash, you'll just assume that they're a prostitute or a Somali pirate."

"Do you see what is happening?" libertarian journalist Lew Rockwell wrote in April 2012. "Simply using cash is enough to get you branded as a potential criminal these days."

The Federal Government now tightly limits how much cash citizens are allowed to carry through border checkpoints into or out of the United States. When you withdraw or deposit $10,000 in cash, your bank reportedly is to notify the government of this transaction, thereby marking you for potential surveillance.

Merely doing transactions with $100 bills, the standard currency of illicit drug dealers, can attract government attention. The law has authorized the asset forfeiture of objects carrying detectable traces of cocaine, however small, which reportedly is the case with as much as 90 percent of U.S. currency notes in circulation. This law has been used to confiscate currency on those grounds.

In Italy, large cash transactions have been banned. In December 2011 Prime Minister Mario Monti proposed prohibiting all cash transactions above $1,300, only a bit more than the average American family's weekly income.

Herding the People

Thus governments are herding their citizens towards a cashless society. Are they doing this for other purposes of their own?

In a cashless future where every transaction leaves a computer history, and governments have open access to those records, cashless advocates dream of simply adding a European-style Value-Added Tax (VAT) or potential American FAIR sales tax, of perhaps 20 percent onto every transaction.

These idealists believe that such a tax could replace most others. This tax would be deducted instantly and automatically during each transaction. Citizens would never again need to worry about tax audits or IRS tax forms; indeed, in this Progressive vision the IRS would largely be made obsolete and disbanded by automatic, unavoidable taxation.

And when government requires more revenue, this tax-on-everybody-all-the-time could instantly be bumped up another percent or two, as Euro-socialist states do with their VAT to close budget shortfalls. What, these cashless-loving Progressives ask, could be more wonderful?

A Tax on Everything

To make such a system work, its advocates quietly add, all transactions must be equal and each one must be a taxable event. Give your son $100 for high school graduation? The 20 percent tax is automatically deducted. Buy an old lamp at a neighbor's yard sale? Pay 20 percent. Lose a sports bet with a co-worker? Pay 20 percent. Give to a hungry poor family? Or to your church or synagogue? Or to the politician demanding a bribe to approve your building permit? Pay 20 percent. No exceptions.

As with the FAIR tax, do not trust the government to curtail all other taxes in exchange for an instant tax on your cashless transactions. The income tax, property tax and others will return to take ever-bigger bites out of your earnings as well. In fact, get ready to pay that extra 20 percent on your transactions paying each of these other taxes.

Progressives are far too ideologically dedicated to soaking the rich to be content with any truly egalitarian tax system.

Much as gold was replaced by inflation-plagued paper fiat money, the cashless society envisioned by such advocates could replace paper fiat money with magical digital money that is always connected to the government – and that loses another 20 percent of its value to the government every time you use it.

Cashlessness as Control

In such a Brave New World, of course, there will be no place to hide. You cannot purchase food or fuel without the transaction being monitored. Your cell phone, by government mandate, contains a locator chip that makes you easy to find (unless you remove its battery).

Could the American Revolution have been won if our Founders lived in such a cashless society? If every transaction must be cleared by a central computer, then whoever controls that computer can monitor or block any transaction a targeted individual attempts to make.

The George Washingtons and Thomas Jeffersons of a future cashless America could easily be neutralized by Progressive redcoats.

Boomer Resistance

"There are a lot of baby boomers who aren't dead yet, and they're simply not going to give up cash," said Ron Shevlin, an analyst with the Aite Group. This is why he sees the near-term future as "the less-cash society, not the cashless society."

Boomers are not the only ones suspicious of spectral ghost money and the attachment Big Government increasingly has to it, according to science and technology writer Glenn Zorpette.

"We're kind of lawless, and we'd rather the government didn't know everything we do," wrote Zorpette in the June 2012 electrical engineering journal *IEEE Spectrum*. "There's a big, spinning world of under-the-table transactions, and what makes it go round is cash."

The Mark

Anyone, whether Christian or not, might find it disquieting that the Bible (Book of Revelation 13:17) foresaw a coming day when only those who carry the mark of the ruler's number on their bodies will be able to "buy and sell."

We are now beginning, as mentioned earlier, to use devices that can read human "biological identifiers," from hand veins to the unique Iris pattern in our eyes.

These devices, using technologies developed by IBM, Hitachi and other companies, then link these identifiers to our Social Security number.

When the Federal Government began issuing these numbers, it promised that they would never be used for any other purpose than Social Security. Today our Social Security number is our government identifier for almost everything. It will inevitably be used as part of our identity in a future cashless society after the last physical dollar has died.

Some U.S. cities, writes Lew Rockwell, have already made it mandatory to implant microchips into all dogs and cats. These same tiny Radio-Frequency ID (RFID) chips could just as easily be injected by needle under human skin.

In the Chips

In 2006 two employees who work at a Cincinnati, Ohio company's secure data center were reportedly implanted with RFID chips.

The journal *Advance for Nurses* reported that in a test begun in 2007, "about 100 patients and caregivers" at an Alzheimer's Disease patient care facility in West Palm Beach, Florida were implanted with RFID chips. This was done so that patients who wandered off and got lost could be identified by other health professionals.

One company called Somark, reported Rockwell, has developed chipless ink that can be applied directly to the skin to create an "RFID tattoo" that can be read by a device from up to four feet away.

RFID chips have also been used in security badges, key chains, and school uniforms. The cell phones used by Google Wallet and Isis also contain a similar chip that allows identification merely by waving the phone near a scanning device.

If paper fiat dollars had a longer life ahead of them, government planners would doubtless consider using RFID ink in printing them. Then our currency, too, could be identified by scanning in yet one more way.

Ego and I.D.

In a cashless Progressive welfare-state future, you may be required to accept whatever the state requires to prevent identity theft.

The cashless future will be more dependent on central identification and control than life was for our independence-loving ancestors.

Your identity, and with it government benefits, may become the closest thing you have to property.

Soon the only place Progressives will not require you to confirm your identity is the voting booth.

Government spending is likely to increase, using spectral money spun out of nothing. High inflation could be routine, and wages will not keep up with it.

As the average American family falls farther and farther behind, home ownership will be replaced by renting. And renting will produce a more transient view of life.

Your car will also be leased, not owned, expanding the economic pattern the government has created by how it shaped tax write-offs.

To repeat the old libertarian saying: government breaks your legs, then offers you crutches and expects you to be grateful.

Progressive government taxes away so much of your earnings that you cannot save enough to retire. Next it offers you a cheap, yet needed, collection of welfare benefits as you work on into your 80s and die in harness.

"Lots and Lots of Digital Zeroes"

If America survives until 2031, when the last Baby Boomers reach full retirement age, more than 65 percent of American households could have someone living there who gets a government check. As noted before, 49.1 percent of American households already did as of 2012.

America's democratic republic will still have elections, although these may become mere bidding wars. The winners will be those who promise to increase government benefits the most. The benefits, however, will be paid in ghost dollars.

Sometime between now and then, if the nation fails to change direction, America will collapse from its own economic contradictions, soaring debt, and loss of values.

What new depth of the modern Great Debasement will come in a cashless society of ghost dollars? A letter-writer to the journal *Technology Review* named Curt Howland foresaw it this way: "When the U.S. dollar collapses, I doubt it will be by printing paper bills. They will simply issue debit cards loaded up with lots and lots of digital zeroes."

"Anyone without one, well, too bad," concludes Howland. "No collectors items this time around, folks, nothing to see here. Move along."

Ghosts in the Machine

The spectral dollar will not be the only ghost haunting the giant computers that house the world economy.

Demons will be there, too, and our brightest hacker minds will fight to prevent their bombs from blowing holes in the walls of our once-proud fortress.

Westerners reportedly directed a cyber weapon known as Stuxnet against the computers controlling Iran's centrifuges to cripple their ability to enrich uranium to atomic bomb-grade nuclear material.

By 2012 a Stuxnet relative code-named Gauss was found stealing banking information in Lebanon and elsewhere in the Middle East, including Israel.

This bug includes a mysterious module called Godel, precise purpose and capability as yet unknown, at least to the public.

Another computer infestation in the Middle East is Flame, a huge digital espionage and malware weapon that the *Washington Post* reported is a joint U.S.-Israeli operation.

Mahdi, named for a messianic figure prophesied by Muhammad to bring the final global triumph of Islam by fire and sword, is of uncertain origin. It appears too unsophisticated to have been cooked up in the computer labs of a government, where devising and defending against cyber warfare and a potential digital Pearl Harbor is the order of the day.

The Center Cannot Hold

By moving towards a cashless society run by computers, we have gambled that these computers will continue to function. If they break down, our unplugged future cashless economy would disintegrate overnight.

Whatever its failings, hard money decentralized America's wealth. When banks failed, people still had gold coins under their mattresses. The frontier was full of independent people who grew their own food, had a well with their own water supply, grew and cut their own firewood, and had the means and skill to shoot their own game...or attackers.

Progressives want to make us dependent on a highly-centralized Big Government that takes away our ability to sustain ourselves, then tries to buy our votes with the stolen fruits earned by others.

Their war against "sprawl" is little more than a tawdry attempt to force those who fled to suburbs and exurbs back into Democrat-controlled cities where they can again be heavily taxed to fund the local welfare state.

(Unlike the Federal Government, states and cities cannot print their own limitless supply of money, which is why profligate Progressive bastions such as California and Illinois will soon become welfare dependents of Federal bailouts.)

Progressives have created a fragile, brittle society controlled from central authority. This makes all of us vulnerable to computer flash-crashes, and to hacker and terrorist attacks on the centralized electronic brain that will soon control our money and economy.

Our Politically Correct schools leave young Americans untrained in how to survive when things break down, skills that our pioneer ancestors knew well. We are taught to be helpless and incompetent precisely to make us dependent, not independent as individuals and a nation.

Welcome to the 21st Century road to serfdom, the road to an emerging Progressive collectivist future.

A collapse of the cashless society and ghost dollar would force Americans to choose a new path to our economic and monetary future.

Will we plunge helplessly into a new Dark Age, or decide to return to something that worked wonderfully – the age of gold that came before the Progressives' Great Debasement and killing of their paper fiat U.S. Dollar?

Chapter Eleven
New Dark Age
or Golden Age?

"You stand in the midst of the greatest achievements
of the greatest productive civilization and you wonder
why it's crumbling around you,
while you're damning its life-blood – money.
You look upon money as the savages did before you,
and you wonder why the jungle is creeping back
to the edge of your cities."

– Ayn Rand, *Atlas Shrugged*

"Time will run back
and fetch the age of gold."

– John Milton

"What comes next for the dollar, Patrick?"

The night's conversation had been long, yet Ryan and Peggy felt surprisingly awake and alive, like explorers on the frontier excited by the new worlds and ideas their economist nephew had revealed.

"Our revels now are ended...melted into air, into thin air," said Patrick, pushing his half-full wine glass aside. Seeing their quizzical looks, he explained: "William Shakespeare, *The Tempest.*"

The Last Bubble

"And like the baseless fabric of this vision....shall dissolve and, like this insubstantial pageant faded, leave not a rack behind," Patrick continued. "In the end the spectral dollar will dissolve into thin air, whence it came."

"How, exactly, will it happen?" asked Ryan.

"In any of a thousand different ways," smiled Patrick. "We know this. There will be the last of many bubbles, inflated with yet another gust of debased stimulus money from the Federal Reserve."

"This is what happened with the real estate bubble, the dot-com bubble, soon the student loan bubble, and so many more – all of which have been caused in our time by distortions in the marketplace from government and Fed interference with the economy."

"The shockwaves caused when this final overvalued bubble bursts will shatter the world's faith in the U.S. Dollar and send them elsewhere in search of a more reliable medium of exchange."

"And will that be the end of the dollar?" murmured Peggy.

"Not quite. The end could be very bad...or ultimately very good. When the dollar's ghost is gone, it will leave not a thing behind, as Shakespeare said."

"It will, however, leave a vacuum, an empty spot where the dollar once held a special place in human minds and hearts, not only in America but also around the world."

More than Money

"The U.S. Dollar was more than money to humankind," said Patrick. "It was a tangible symbol of the United States – a land consecrated to human freedom and opportunity."

"The dollar," explained Patrick, "was the green energy of the G.I.s who selflessly died to liberate Europe and other lands."

"The dollar fueled the pioneering spirit that made Neil Armstrong's the first human footprint on the Moon."

"The dollar was the coin of the realm on both the old and the new frontier of human dreams, from covered wagons to the stars."

"The dollar, so hated by Progressives, was the great equalizer through which anyone with luck, pluck and talent could succeed in America – regardless of race, creed or status at birth," Patrick went on.

"The dollar did not make us unequal, as government-loving Progressives claim. The dollar made all who earned them equal...something mere money could not do in the aristocratic class systems of Europe."

"The dollar was to the world the tangible symbol of American generosity, of our innocent idealism, and of our prosperity in a land of all lands where people from everywhere could come together, take part in free enterprise and succeed."

"As President Ronald Reagan said: You could move to Japan but they would never regard you as Japanese; yet anyone can legally emigrate to the United States and be welcomed as an American."

Stepping Towards a Dark Age

After thousands of years of worldwide racial and national conflicts, the U.S. Dollar emerged as the global symbol of a land that is a successful model for the future of a free planet united by traditional American values and ideals.

"This is what those Progressive fools killed when they kept debasing and politicizing the dollar," said Patrick, with a catch in his throat. "They destroyed humankind's best way and means and hope of a better future. What may soon come instead is a new Dark Age."

Historians may someday mark September 13, 2012, as the date of the final fatal step off the cliff for the U.S. Dollar.

On that unlucky date, Federal Reserve Chairman Ben Bernanke announced that the Fed would be providing economic stimulus of $40 Billion every month – which over a year would be nearly half a trillion dollars, similar to his previous quantitative easings under QE1 and QE2.

Bernanke's new stimulus, however, would be "open-ended," with no end date whatsoever. It was simultaneously a signal of two seemingly contradictory things.

It promised speculators that the drug pusher at the Federal Reserve would keep supplying their "fix" of nearly-free stimulus money from thin air with which to place low-risk bets.

To serious investors who actually want to build companies and hire people, on the other hand, Bernanke's message was that this monthly stimulus money could suddenly stop at any time – and that, when the Fed cut it off, our addicted economy would plunge into addiction withdrawal symptoms that could crash the market and destroy their investments. The contractions and seizures of this withdrawal would bring our addicted economy to a screeching halt.

QE3's Fatal Trap

Both speculators and investors understood the down side of Bernanke's stimulus policy and politics. It would produce a sugar high, a hallucination of false prosperity, if it worked at all.

The drug of QE3 is not the cure for the economy's problem. To paraphrase President Ronald Reagan, Quantitative Easing 1, 2, 3 and beyond *is* the problem.

Knowing what Bernanke has promised, the market has already discounted the Fed's $40 Billion per month, thereby nearly erasing the ability of this credit infusion to stimulate real growth or more jobs.

The Stimulus Drug

QE3, however, like other addictive drugs hooks the economy on needing ever-more and ever-larger doses of stimulus just to stay where it is.

Welcome to the American economy's "new normal" – a form of perpetual stimulation economics perfectly suited to a culture where people use more pain-relieving, mood-adjusting, passion-reviving and mind-altering drugs than any other society in recent human history.

Welcome to lotusland, where even our money is a drug.

As we explained, documented, and quoted experts such as Nobel-laureate economist Milton Friedman warning in our book *The Inflation Deception*, this kind of deliberate inflation of the currency first affects users with the hypnotic illusion of prosperity; it quickly leads to addiction and crashes the user with a devastating hangover and severe economic sickness, even death of the currency.

Bernanke's Put-Put

Long before Chairman Bernanke spoke on September 13, 2012, everyone knew that the Fed and Federal Government were trapped. They had little choice, especially with a national election approaching, except to keep injecting more and more debased fiat money into the marketplace, just to prevent a collapse.

They could not take the money back via higher taxes, as even Keynesian theory called for, without sucking the capital out of capitalist business investment, hiring and productivity.

Such taxation in a down economy, economist John Maynard Keynes knew, would damage an already-bad situation. Every dollar taken in taxes, economists David and Christina Romer calculated in 2010, could prevent $3 in economic growth per year for years to come.

Yet if the Fed or Federal Government did not retrieve the trillions already injected as stimulus spending, future soaring inflation would inevitably crash the dollar and sink the economy as everything denominated in dollars became worthless, like the money in Weimar, Germany, after World War I.

A New Dark Age

In 1932 a reporter asked John Maynard Keynes if there had ever been anything like the Great Depression.

"Yes," replied Keynes. "It was called the Dark Ages and it lasted 400 years."

As monetary journalist Ralph Benko in September 2012 noted in recalling Keynes' remark, we are already living in a kind of "Little Dark Age."

"America is entering its fifth decade of punk (less than 3% average annual GDP) growth and second decade of pure stagnancy – growth arguably averaging under 2%," wrote Benko. "Less than 3% is bad, but less than 2% is slower than population growth. That implies a severe absence of opportunity to flourish: a 'Little Dark Age.'"

During the Second Quarter of 2012, the U.S. economy was growing at 1.3 percent (or minus 5 percent if our real rate of more than 7 percent inflation is subtracted, which would show that the Great Recession is still here).

The late longshoreman philosopher and author Eric Hoffer once observed that if you ask a leftist or Progressive intellectual what time in history he would like to live in, the answer you will most often get is the Middle Ages. One of us once asked the famed Marxist psychoanalyst, Erich Fromm, Hoffer's question, and got exactly the response predicted.

Why do Progressives love this pre-Renaissance, pre-Enlightenment age? Because, said Hoffer, this was the last time in Western history that intellectuals (in the form of theologians) were routinely part of the elite that ruled the state.

Perhaps Progressives subconsciously want to create a new Dark Age because this is what societies turn into when ruled by intellectuals. It is their natural environment.

Progressives have already moved the world towards a Dark Age by undermining the Enlightenment philosophies of free enterprise, free speech and free thought that were embodied in the American Revolution.

And like history's Dark Ages, today's stagnant Progressive economy has its own plague, its own New Black Death, wrote *Forbes* Magazine columnist Louis Woodhill in May 2012. This "plague is the result of Keynesianism," whose ideas have pushed Europe to the brink of economic collapse, and if Progressives get their way could do likewise here.

"In the 1300s, the Black Death killed about a third of Europe's population," wrote Woodhill. "If its spread is not checked, Keynesianism may wipe out a third of Europe's GDP (Gross Domestic Product)."

Easy Money

The Fed and the Feds have conjured a completely artificial economy, a deceptive and false illusion of growth based on unreal money spun from nothing, and produced a constant sugar high that pushed the economy into financial diabetes.

This situation is, in the words of even President Obama's Treasury Secretary Timothy Geithner, "unsustainable."

Ultimately people and economies to survive must produce more than they consume.

Governments cannot for very long grow faster than the economy that supports them.

Growing Government

Yet the Progressives have built exactly that, a system that is devouring its seed corn and strangling America's producers.

President Obama in less than four years increased the size of government from less than 20 percent of America's Gross Domestic Product to nearly 25 percent of GDP, enlarging the Federal Government's piece of our economic pie by roughly 25 percent, a huge expansion of government in a stunningly short time.

At a time of sky-high private sector unemployment, caused largely by government economic policies, Mr. Obama increased employees of the Federal Government by 11.4 percent. As we noted before, each of these employees, on average, is paid $126,000 a year in wages plus benefits, roughly double the earnings of private sector employee-taxpayers.

America's Founders were familiar with this kind of Big Government. In the Declaration of Independence, Thomas Jefferson wrote that King George III had "erected a Multitude of new Offices, and sent hither Swarms of Officers to harrass our People, and eat out their Substance."

Borrowed Money

Today's huge expansion of the Federal Government is being done not only through higher taxes but also through the cruelest tax of all, interest and inflation caused by massive borrowing.

Roughly 41 cents of every dollar the Federal Government is spending is borrowed...and under President Obama this borrowing reached $58,000 every second.

This has made inflation America's biggest export. The People's Republic of China and Japan each have bought more than $1.1 Trillion of America's debt obligations, and both expect to be repaid on their U.S. Treasury obligations with interest.

In 2011 total U.S. interest payments on America's immediate $16 Trillion debt were close to half a trillion dollars. They will go stratospheric if the Fed is unable to hold rates down, as will happen if credit rating agencies further downgrade the credit worthiness of America's "full faith and credit," the only thing that ultimately our dollar stands or falls on.

At some point China and/or Japan, seeing that the U.S. has little prospect of ever paying off our debts with dollars of anything like the same value they used to purchase our bonds, will demand a higher rate of interest.

Such demands might not even be vindictive or political. Japan's economy has been hobbled by earthquake, tsunami and nuclear reactor disasters. China's economy, after 30 years of purported growth, may now be in serious contraction. Both might simply need the money to support their own economic health.

When will this happen? The answer is beyond America's control. Nearly half of the Federal Government's debt paper will come due, and require refinancing, over the next four years. Although the Federal Reserve itself now buys as much as 70 percent of U.S. debt obligations, China and Japan and others will have a significant say in how expensive this debt on tax-payer shoulders becomes.

Rollover

Such a rollover of massive U.S. debt – even in the relatively short run, $16 Trillion and rising, which in 2012 surpassed America's entire annual Gross Domestic Product – could add literally trillions of dollars in annual interest charges to what our spendaholic politicians have put on America's government credit card.

This could be the straw that breaks the economy's, and hence the dollar's back.

Something called the U.S. Dollar would remain. With America's economy being devoured by a black hole of debt, however, the dollar would not long remain the world's Reserve Currency. It would either lose this status or, more likely, be forced to compete for this lucrative status with one or more other currencies.

As suggested earlier, one contender to be the new global Reserve Currency would be the Yuan, aka Reminbi, the currency of China, now the world's #2 economy. Another would be the Euro, if it survives. A third, at least in theory, would be a new scrip based on a basket of global currencies such as the International Monetary Fund (IMF) Special Drawing Rights exchange medium. A fourth would be an attempt to create a United Nations currency modeled on John Maynard Keynes' notion of the Bancor.

More likely, however, would be a global economic breakdown that might prompt countries, and even multinational corporations, to turn to small national and corporation exchange methods or computerized barter.

The United States, Canada and Mexico during the late 1990s considered a regional common market currency to be called the Amero. In 1999, Herbert Grubel of Canada's prestigious Fraser Institute did a 50-page study of this idea titled *The Case for the Amero* which analyzed the pros and cons of a North American Monetary Union akin to the Euro Community. This controversial idea was associated with then-Mexican President Vicente

Fox and has been little discussed since he left office.

The United States and Canada, with their similar advanced economies and dollars already at near parity, could probably agree to use a common currency. Using the same currency when countries are economically as different as the U.S. and Mexico would likely be more problematic, as Euro Community issues between Germany and Greece have shown.

Worth noting: As of 2011, Mexico had the 14th largest annual Gross Domestic Product in the world – larger than the GDP of South Korea, the Netherlands, Switzerland, or even Saudi Arabia. Mexico's GDP is larger than any other Latin American country's, except Brazil's. Mexico's GDP is approximately two-thirds the size of the #10 GDP country, Canada, whose population of 34.5 million is only 30 percent of Mexico's.

Buffer Zones

If the dollar collapses as the global Reserve Currency, much of the global economic system could also begin falling apart.

In the dollar's wake, the world could become a mix of heavily-armed city-states and nations. China and Russia signed an agreement in November 2010 to trade with one another using their own national currencies, not dollars.

Russia, however, has continued a policy from several decades ago. It provides lucrative mining concessions to Japanese enterprises in regions along the Russo-Chinese border. Why?

In the political chess of geopolitics, Russia knows that China hungrily eyes the vast riches of oil, rare earths, gold, timber and other natural resources Moscow controls in Siberia.

In response, Russia welcomes Japanese companies to become a buffer zone, to position thousands of Japanese workers where they would be injured or killed if China invades Russia. These Japanese are, in effect, volunteer human shields – tripwires whose deaths would put great internal political pressure on Japan's government to side with Russia once such a war began. Japan seems unconcerned; the likelihood of such a Chinese invasion seems vanishingly small.

Russia is a militarily-powerful nation, yet also until recently an empire that is now bordered by many of its former colonies. It is a treasure trove whose government is increasingly close to the Christian Russian Orthodox Church.

Mindful of Russia's wealth and declining Great Russian population are China to its southeast, a swath of Muslim former Soviet states along its southern border, and Germany to its west.

Russia has chosen to play an odd game, helping Iran acquire nuclear reactors and advanced atomic technology. The weapons they today are helping this Muslim theocratic state acquire could someday soon be used against them by Islamist terrorists.

In a chaotic world, Russia could quickly become a fortress under siege.

OPEC and OFEC

In a disintegrating world economy without a reliable Reserve Currency, some blocs may return to some kind of barter. OPEC, the Organization of Petroleum Exporting Countries, could simply issue its own currency, the OPECy, worth one barrel of light sweet Saudi Arabian crude oil. Harder-to-refine oils such as Mexico's tar-heavy crude or America's paraffin-rich North Slope Alaskan oil would have a slightly lower OPECy exchange value.

As a new economic Dark Age descends, the United States could likewise turn its resources into a new barter currency.

The U.S. could launch OFEC, the Organization of Food Exporting Countries. OFEC would strive not to exploit famines or starvation, yet like OPEC it could use its resource as a weapon to press for political alliances and increased food income for its members.

OFEC would not include some of the world's biggest growers of wheat, rice, corn (maize) or millet – China and India – because they need these crops to feed their own huge and growing populations.

The U.S. – the world's biggest grower of corn and sorghum and third biggest of wheat – traditionally grows four times more food than it consumes.

This has changed a bit with 40 percent of America's corn crop going to make ethanol fuel for its vehicles.

Potential U.S. partners in OFEC could be Canada, Australia, France, Germany, Poland, Ukraine, Brazil, Argentina, and Russia.

Turning American plowshares into double-edged swords has long been debated.

In 1980 President Jimmy Carter cut off U.S. grain shipments to the Soviet Union in response to its invasion of Afghanistan. Some analysts later argued that this proved how ineffective the "food weapon" can be. To be fair, it is difficult to successfully use food as a weapon against one of the world's largest food producers.

The aim of OFEC would not be to starve nations to their knees. On the contrary, OFEC might provide free food to nations suffering flood and droughts, as the U.S. has done for more than a century. Our experience, alas, is that when we give boatloads of free food to poor nations, it can leave their own farmers unable to earn a living and puts them out of business.

As with Progressive welfare here at home, the most cynical use of food as a weapon would be free food aid that leaves nations no longer able to feed themselves, and hence dependent on outside help. Such food aid in the past gave the United States leverage with Egypt to encourage a peace agreement with Israel. That peace agreement now appears at risk of unravelling.

OFEC could also give its members a bit of leverage against OPEC. In anticipation of future global food shortages caused by Earth's weather returning to a more erratic normal, Saudi Arabia and other oil-rich nations have over the past decade bought vast amounts of prime farmland in Africa, North America and elsewhere.

The New Feudalism

In this new age of feudalism, alliances that use commodities as money such as OPEC and OFEC should also brace for the emergence of what could be called OTEC, the Organization of Terrorism Exporting Countries.

More than a decade after 9-11, we are already deep into an age of terrorism whose most dramatic act was the symbolic economic destruction of western civilization's World Trade Center in New York City, aimed directly at the heart of Western capitalism.

Technological change alters how we structure our societies. As the late media theorist Marshall McLuhan was fond of saying, we shape our tools and thereafter our tools shape us. Look how our lives have been re-shaped because Progressives forced paper fiat money on us.

The old feudalism was built around stone castles, which serfs and knights and dukes depended on for protection in a Dark Age of marauding bands.

Gunpowder and cannon made stone castles obsolete.

In the 20th Century we built our enlightened civilization on far-flung trade, shopping malls, ever-more centralized technology, and debased paper fiat money. This civilization now teeters on the edge of an economic cliff.

What happens when high-tech terrorism becomes the new gunpowder, and single attackers can create mushroom clouds and radioactive holes in places we used to call Phoenix or Dallas, Philadelphia or Miami, Las Vegas or Chicago, Detroit or Portland, Boston or San Diego?

What happens when the biggest weapons are the smallest – mycotoxins, anthrax, or the "human insecticide" VX nerve gas that can be made at a cost of $5 a quart...and a quart can potentially kill a million people in a crowded big city?

What happens when tiny bands of terrorists armed with gigantic weapons from Iran or other foes begin systematically to target our liberty, computers, and ability to travel, shop and trade freely?

What happens when such terrorists acquire the ability to use degaussing cannons and electromagnetic pulse (EMP) weapons, one type of which we discussed earlier, to wipe out the computer data and circuits that run our cities, keep our cell phones and cars working, control our national power grid, and give reality to the ghost money in our government and banks on which our way of life depends?

What would you do if such assaults unplug our computers and telephones, and your ghost money and credit vanishes? Or when you try to withdraw $100 at your bank, only to be told that their computers show a zero balance in your account? Or when the dollar collapses, and nobody will accept your pieces of green paper anymore because nobody knows how much or little they are really worth?

This could not have happened in the years just before the Great Debasement began in 1913, because the money jingling in your pocket or purse had intrinsic precious-metal value that was known and understood by everyone.

Are you ready to live in new castles whose defenses include massive surveillance and a complete loss of individual privacy?

Are you ready in this fragile global economy on the brink to become a serf or technopeasant in the new feudal system?

When acts of terrorism appear to justify such loss of liberty, will you wonder like the author of the dystopian novel *1984*, George Orwell, if the government itself is merely faking these events to cement its own totalitarian power?

When the Federal Reserve offers QE-27 on the same week that Apple releases iPhone32 with yet more ways to spend your government welfare stipend, will you wonder like Aldous Huxley, author of the dystopian novel *Brave New World* (whose title also came from Shakespeare), if any of this is real – or merely a hallucination induced by the drug Soma or by government stimulus?

The Tytler Cycle

A social breakdown may be the catalyst needed to make us look within and rediscover the values that originally turned the United States into the greatest hope of humankind.

As we discussed in *The Inflation Deception*, America may be nearing the low point in a long cycle civilizations experience. This cycle was described by Scottish historian Alexander Tytler, who wrote of it at the same time as the American Revolution.

The cycle begins with a people in bondage, wrote Tytler. From bondage they turn to spiritual faith, which inspires them to courage, which moves them to take actions that win their liberty.

As the cycle continues, liberty and faith and courage guide this people to achieve material abundance.

What then has happened, over and over throughout human history, wrote Tytler, is that abundance leads to selfishness, to feeling entitled to consume more than they produce, and to taking their freedom and success for granted.

As the downward cycle continues, selfishness begets complacency and apathy as people forget the values, work ethic and spiritual faith that led them to abundance and liberty.

Again and again, civilizations have tried to sustain their lifestyle and material wealth by going deep into debt in one way or another. This dependence soon returns a people to bondage, where centuries or millennia may pass before a new leader or generation finds the spiritual faith to begin the upward cycle anew.

It is not hard to see where many Americans are on Tytler's cycle. The challenge before us is to be a saving remnant, able to restore the uplifting values that move a people to greatness and renew the vision of America's Founders – for where there is no vision the people perish.

Or are we content to take the money and run, to take the credit that now pretends to be money, and let real cash go?

Once we lived in a vastly freer republic that was made possible by a very different kind of money.

A Different Past, a Different Future

People are beginning to ask: what if we and the other industrial nations had never abandoned the gold standard that once worked so well?

What if Progressive, self-serving politicians had never created the Federal Reserve and the Great Debasement that has destroyed America's Golden Age?

World War I then might never have happened, because none of the nations involved had enough gold to pay for such a war. Even if this "War to End All Wars" had occurred, America would likely have stayed out of it and pressed both sides to stop the fighting far sooner.

To help it win World War I, Germany transported Marxist firebrand Vladimir Lenin to the edge of Russia – thereby helping ignite the spark that brought down Russia's democratic Kerensky government (which six months before its own fall had replaced the Czar).

Without World War I, in other words, Russia would never have lost millions of soldiers and been brought to the brink of Marxist revolution.

The 20th Century would have had no Marxist Soviet Union, and hence no later Communist revolution in China, no Cold War, no American wars against Communist forces in Korea or Vietnam, and no cultural revolution of the 1960s that warped today's Progressive politics of class warfare and the American welfare state.

A History without Hitler

Without World War I, no Weimar hyperinflation would have happened. Its destruction of German money's value also destroyed the moral values of hard work and thrift, as we explored and documented in our books *Crashing the Dollar* and *The Inflation Deception*.

Those who saved their evaporating money were suckers during the Weimar hyperinflation. Those who got rich were the same sort that Ben Bernanke's "easy money" Federal Reserve is enriching today – speculators and gamblers, who bought real, tangible assets and then repaid their debts with debased, worthless money.

This was the same problem with paper money that Thomas Jefferson and George Washington understood so well. This is why they wanted the U.S. Dollar based not on paper but on precious metal that politicians and central bankers could not, without limit, run off a printing press.

This economic demolition of traditional values in the Weimar hyperinflation paved the way for those who promised giant public works projects and restored greatness, "Progressives" like National Socialist (the German contraction of which is "Nazi") politician Adolf Hitler.

Today's Progressive government schools predictably never teach their students that Adolf Hitler was a socialist who despised free markets, preached environmentalism, imposed gun control, favored outlawing the smoking of tobacco, and was a vegetarian.

World War II was a continuation of World War I, so if World War I never happened, then Hitler's Third Reich and the Holocaust might never have happened.

The World We Lost

How different the world would be if we never let the Progressives take away our gold-backed economy and currency.

We know what has happened since 1913. What few of us were taught in the government schools of Progressive Education was what we lost, what might have been had America not been steered off the yellow brick road of solid money.

We lost an honest world already united by the common gold basis of major nations' currencies. This fostered trade and friendship among nations.

We lost a world where money kept its value so that workers were not robbed by inflation or income taxes of the fruits of their hard work and thrift, just as Thomas Jefferson promised.

We lost a world in which, so long as nations honored the gold standard, war could not be fought on a credit card. The weapons of war had to be bought with hard national savings, which meant that politicians had to think long and hard before starting a war. This made peace and prosperity more likely.

We lost a world where Americans were free.

"Freedom is never more than one generation away from extinction," said Ronald Reagan in a 1961 Phoenix speech. "We didn't pass it on to our children in the bloodstream. It must be fought for, protected, and handed on for them to do the same, or one day we will spend our sunset years telling our children and our children's children what it was once like in the United States where men were free."

The Greatest Debasement

"The Greatest Debasement has not just been the devaluing of our money," said Patrick.

"The Progressives have also debased America's morals, our bedrock values of hard work and thrift and integrity, our spirit of individual responsibility, our sense of fairness and justice, and our belief that achievement deserves to be rewarded."

"What the Progressives have really debased is America itself."

The Fork in Our Road

President Ronald Reagan appointed a Gold Commission to see if we could restore that world of sound money, dependable values, and almost zero inflation. One of this Commission's members was Texas Congressman Ron Paul, who ever since has known that gold could be the other path that gets us off the collectivist Progressive path that Nobel laureate Friedrich Hayek called "the road to serfdom."

The Republican Party in 2012 included in its national platform a call to create a new Gold Commission to consider whether to base our money again on the U.S. Constitution ideal of our Founders – precious metal-backed money that puts golden handcuffs on our spendaholic politicians.

The Federal Government simply dares not begin printing limitless amounts of gold-backed dollars. If people suddenly saw more dollars in the marketplace, they would be able to demand that the government exchange their paper Gold Certificates for actual gold.

Rep. Paul, who in 2012 turned 77, announced that same year that he would not seek another term in Congress.

Despite Congressman Paul's position as Chairman of the House Subcommittee on Domestic Monetary Policy and Technology, the Federal Reserve has refused his repeated requests for an independent audit of how much gold is actually in the U.S. gold reserve in Fort Knox, Kentucky.

Return to a Golden Age

As we mentioned in Chapter Eight, China is afraid. As Jon Huntsman, our Ambassador to China in 2010 spelled out in a State Department cable, China fears that the U.S. will suddenly devalue the U.S. Dollar and then immediately peg our currency to gold.

In 2009, investment banker and Pentagon consultant James Rickards was one of several experts who took part in a "war game" to evaluate the risks America might face from international financial warfare.

Rickards in this game put forth the following idea: what if Russia unexpectedly set up a new bank in London and launched a new currency backed by its Swiss bank gold reserves and position as one of the world's top gold producers?

What if Russia announced that its oil, natural gas and other international sales now be transacted using only its new currency, not U.S. Dollars? Could it displace the dollar and become the world's new Reserve Currency?

Rickards describes what happened during this financial war game in his 2011 book *Currency Wars: The Making of the Next Global Crisis*.

"[W]e're tired of the U.S. using its dominant position in the dollar-based trading system to call the shots. There's a better way," one of the strategists told Rickards. "None of our [paper fiat] currencies is ready to replace the dollar – we all know that."

"But gold has always been money good," the strategist continued with an odd twist of words. "It's just a matter of time before the world gets to some kind of gold standard. There's a huge first-mover advantage here. The first country that moves to gold will have the only currency anyone wants."

As Rickards quickly discovered, the others in this war game had no good way to counter the winning advantage of a new Russian gold-standard currency and its ability to displace the dollar in world trade. Their response was simply to prohibit his use of this gambit in the war game.

We believe Rickards is right – that whatever country makes available a new gold-backed money will gain tremendous advantage in a world of debased, declining, distrusted fiat paper currencies.

That nation will win over the dollar and other nations' paper that have been competing in years of currency wars by seeing which could make their currency the least valuable, and hence their exports the cheapest.

We have been in a paper fiat money race to the bottom, a "race to debase," so it should not surprise us that the major fiat currencies such as the dollar are now low and getting lower in value.

Our exports may be cheap in foreign markets, but we no longer have the purchasing power to afford the imports or commodities we want. In trying to beggar our neighbor, we have beggared ourselves.

A New Denarius?

The Dark Ages arrived when barbarians overthrew the Roman Empire, which had already been hollowed out by its own Great Debasement, by "Progressive" emperors who made its once-reliable money worthless and raised taxes so high that many Romans greeted the barbarians almost as liberators. We explored the history of this in *The Inflation Deception*.

However, only half of the Roman Empire fell in 476 A.D. The Eastern half, known as the Byzantine Empire, survived and mostly prospered until the Ottoman Turks overthrew it in 1453 A.D., less than 50 years before Columbus reached the New World.

Byzantium survived, in large part, because its money remained solid...so solid that it was named the Solidus, the gold coin used to pay its troops, who from it came to be called "soldiers."

This reliable gold coin became the symbol of the empire's strength and integrity across the Byzantine Empire and in lands beyond.

Successive Islamic Caliphates, to demonstrate their own merit, minted gold coins that resembled Byzantium's. Both coins became a standard for trade in much of the world and together were known as Bezants.

The Ottoman Empire conquered the Byzantine capital Constantinople, thereafter seized Greece and parts of Eastern Europe, but in decadence declined – in part because of their own Great Debasement, as we soon shall see – and was overthrown in 1924. When the Ottoman Caliphate died, so did its signature coin named for the ancient Roman Denarius, the Islamic Dinar.

Efforts have been made since then to re-establish this Islamic Dinar as a global gold money that could replace the U.S. Dollar as the world's Reserve Currency, first as the medium in which Muslim nations sell their oil, and then in all world trade.

Malaysia's northern state of Kelantan aimed to mint an Islamic Dinar that would be used to pay a quarter of government worker salaries, be accepted by state companies, and become a coin of private commerce.

Writing in 2010 in the socialist British newspaper *The Guardian*, Nazry Bahrawi reported that Islamic Dinar supporters were promoting it as both an antidote to Western capitalists cheating small nations with inflated, untrustworthy paper money, and as a Muslim-unifying religious coin fit to use when paying Zakat, the 2.5 percent-of-their-wealth Muslims are religiously required to give to the poor, others in need....or jihad.

Bahrawi was skeptical, noting that Kelantan's Islamic Dinars were being used both as commodities that profited money-changers, and as political tools by the local ruling political party to advance its power.

The Islamic Dinar has yet to catch on worldwide, yet it has bullion – not legal currency – versions being made in Muslim Indonesia, minted in Bahrain, and manufactured in the United States. The further weakening of the U.S. Dollar might soon trigger greater enthusiasm for this Islamic gold coin.

When first issued in 2001, the gold in Kelantan's Islamic Dinar had an exchange value of $42.50. In 2012 the gold in that Dinar was worth $257. If as an oil producer in 2001 you had been paid in Islamic Dinars instead of dollars, you would, over time, have been far better off accepting the Dinars.

The Once and Future Gold Dollar

Imagine what could happen if the United States in 2013 used the 100th Anniversary of the Progressive's Great Debasement to announce that we have seen the error of our ways and are reversing our self-destructive Progressive policies.

The U.S. Dollar, we should announce, is immediately returning to a gold standard.

Our heavy Progressive taxes will all be rolled back, turning the United States into the friendliest personal and business tax haven on Earth. Within 24 hours trillions in idled investment cash from all over the world will be flooding into the United States, the new home of global entrepreneurship, enterprise and jobs.

Most of us have forgotten, or were never taught in Progressive government schools, that this is what the young United States used to be like, a land where millions came to have the opportunities and happiness they could not find under the political or economic oppression of their own countries.

Our extreme Progressive anti-business and anti-private property regulations and restrictions will immediately be rolled back, an announcement that will spark a boom of investment and prosperity once again in America.

America can be young and vital again, as soon as we clear the artery-clogging, heart-hardening poison of Progressivism from our body politic.

We have only our chains to lose. We have a bright, prosperous future to win for our children and grandchildren.

Is a New Gold Standard Feasible?

One of the reasons Rep. Paul wants an audit of the Federal Reserve's gold in Fort Knox and elsewhere is to help determine whether the United States still has enough gold to again put the U.S. Dollar on a gold standard.

According to Nathan Lewis, author of the book *Gold: The Once and Future Money*, it should be feasible to transition to a gold standard with gold valued at $1,600 per (Troy) ounce, significantly below the level at which gold was trading in September 2012.

Other experts see no difficulty establishing a new gold standard for the U.S. Dollar with gold convertibility at $5,000 to the ounce.

A new gold standard will become law within five years, predicts economic expert and longtime publisher of the business magazine *Forbes,* Steve Forbes.

Why? Because this may be the only way to restore the U.S. Dollar and return America's economy to prosperity.

This will produce winners and losers. Among the losers will be Progressives who were using the Federal Reserve, the Great Debasement, and massive printing of paper fiat dollars to empower themselves and their ideology.

Nearly every other American will win in the short or long run by returning to honest, solid money.

Among the biggest winners could be those who bought gold below its new Gold Standard price, which could be as high as $5,000 to $10,000 per ounce. Those who bought in September 2012 at around $1,750 per ounce, for example, might see a $3,250 per ounce increase in value and buying power if the new standard makes the U.S. Dollar officially worth one-five-thousandth of an ounce of gold.

The best way to think of gold is not primarily as an investment for profit, like stocks, but as an insurance policy that could help protect you against future problems and further debasement of the dollar.

In *The Inflation Deception*, we explained why returning to a gold standard is urgently important.

Gold versus The Welfare State

A silver bullet is widely believed to have the power to kill a werewolf, if such a creature exists.

Gold is the bullet that, without doubt, can kill the very real monster of inflation, which is almost entirely a creature of government paper fiat currency.

This is why in 1966 economist Alan Greenspan, then a member of Ayn Rand's inner circle (yet later, ironically, to chair the Federal Reserve Board) wrote about why so many "welfare-state advocates" have "An almost hysterical antagonism toward the gold standard."

Anchoring a currency to gold convertibility, he wrote, handcuffs politicians who are eager to run the Mint's printing presses.

Welfare-statists, Greenspan wrote, realize that "the gold standard is incompatible with chronic deficit-spending (the hallmark of the welfare state).... [T]he welfare state is nothing more than a mechanism by which governments confiscate the wealth of the productive members of a society to support a wide variety of welfare schemes."

"In the absence of the gold standard," wrote Greenspan, "there is no way to protect savings from confiscation through inflation. There is no safe store of value.... The financial policy of the welfare state requires that there be no way for the owners of wealth to protect themselves."

"Deficit spending is simply a scheme for the 'hidden' confiscation of wealth," Greenspan continued. "Gold stands in the way of the insidious process. It stands as a protector of property rights. If one grasps this, one has no difficulty in understanding the statists' antagonism toward the gold standard."

D.I.Y. Gold

Progressive politicians and their comrades in the media will do all within their means to prevent a new gold standard. The reason is obvious. Gold trumps red. A restored gold standard would be the end of their game.

A restored gold standard would stop dead in its tracks the Progressives' plan to keep expanding government, and would begin trimming our obese, hyperactive Uncle Sugar back towards the lean, size-limited servant of the People that America's Founders intended our republic's government to be.

If the Progressives' frantic resistance to restoring the gold standard to the U.S. Dollar delays this from happening, it need not delay you.

You can create your own gold standard simply by transferring some of the dollars you have saved into what many economists already call the real new global Reserve Currency, the world's once and future legitimate money – gold.

The politicians can, and soon must when their credit is cut off, do what economists call "monetizing the debt."

They will simply turn on the Mint's printing presses – or, with today's digital ghost dollars, simply begin adding "lots and lots of zeroes" to the government's existing computer accounts – and out of nothing instantly create hundreds of trillions of dollars.

Lawless Tender

The government, remember, has the power under the legal tender law to compel creditors who are owed debt denominated in U.S. Dollars to accept this otherwise-worthless paper fiat money as payment in full.

Doing this will, of course, crash the dollar and shatter the full faith and credit of the United States. Today's Progressive politicians know this and, in our opinion, frankly seem not to care.

This is what they are already doing, only a bit more slowly to prevent people like you from rushing to convert your paper dollars into something they cannot run off a printing press in Washington, D.C.

To escape the tender law and their world of ghost dollars and Zombie Euros – you need to free some of your savings and dollar-based investments from the world of government dollars and return to money that is real, the money that America's Founders put into the Constitution and an even higher authority set as the definition of honest money in the Bible.

Engoldenment

Patrick was smiling.

"You'll remember that I told you how confronting a Progressive with the gold standard is like holding up a crucifix to a vampire," he said.

"One other thing stops vampires, at least outside the 'Twilight' saga, and that is the sunlight of a new day."

Patrick, Peggy and Ryan had talked all night. Now, through the dining room's beveled east window, bright rays of sunlight streamed. The room was suddenly filled with rainbows.

"This is the real choice each person has to make," said Patrick.

"We will either have a new Dark Age caused by the Progressives' Great Debasement of disintegrating paper dollars. Or we will have the enlightenment of a new Golden Age – I call it 'Engoldenment,' the opposite of dependence on government and its serfdom money – that opens people's eyes so they can see clearly and rediscover their freedom."

"We choose freedom," said Peggy. Ryan took her hand.

Epilogue
The Ottoman Mirror

"There is no means of avoiding the final collapse
of a boom brought about by credit expansion.
The alternative is only whether the crisis should come
sooner as a result of a voluntary abandonment
of further credit expansion, or later as a
final total catastrophe of the currency involved."

– Ludwig von Mises
Austrian Economist

2013 marks the 100th Anniversary of America's Great Debasement. It began with the creation of the Federal Reserve System and the income tax. In one century, this debasement of the money has shrunk the U.S. Dollar to only two pennies of its 1913 purchasing power.

America's spending now exceeds its income and is, nearly all economists agree, unsustainable. Our debt is so immense that it is unpayable. What does the future hold next for America?

For what could be a glimpse of our future, take a deep look, if you dare, into the Ottoman mirror.

Between 1808 and 1844 the mighty Ottoman Empire had its own "Great Debasement," as Turkish economic historian Sevket Pamuk calls it.

The Ottoman ruler Sultan Mahmud II by 1822 had in stages reduced the amount of silver in the empire's mainstay coinage by 60 percent. Between 1828 and 1831, he reduced the silver that remained in already-debased coins by another 79 percent.

In 23 years, the silver content of the Ottoman Kurus coin was reduced from 5.9 grams to only 0.53 grams.

In 1808, 19 of these Ottoman coins could buy one British Pound. By 1844, a British Pound cost 110 Kurus. Against the leading European currencies, the Kurus, according to Pamuk, lost roughly 83 percent of its value.

Food prices quadrupled, wrote Pamuk, a Professor of Economics at Bogazici University in Istanbul and author of *A Monetary History of the Ottoman Empire.*

Who Benefits from Debasement?

Exactly as with America's current Great Debasement, many suffered and some benefited. The poor, those on fixed incomes, and lenders who wanted to be repaid in money with the same purchasing power they lent, were hurt by the deliberate devaluation of the money and the inflation this unleashed.

Great Debasements' biggest beneficiaries are speculators, those who owned long-term fixed debts they could repay with devalued money, and, above all, the government itself.

Pamuk finds that the government increased its revenue by an average of 10 percent a year by using its Great Debasement as a substitute for increasing taxes....even though much of that revenue was soon lost again to the higher prices its inflation caused.

Like the Progressives behind America's Great Debasement, the Ottoman Sultan debased the money quite deliberately, knowing that this would cause a major redistribution of wealth in society that in the short run would enrich the government.

Silver rather than gold coins were the target for the Sultan's Great Debasement, writes Pamuk, "because the obligations of the state were expressed in terms of the silver Kurus and not linked to any gold coin. As a result, the government did not stand to gain much from debasing the gold coins."

Gold Meddling

In America's Great Debasement, by contrast, in 1933 the Progressive President Franklin Delano Roosevelt made it a crime punishable by prison time and huge fines for an unauthorized American citizen to possess gold bullion. Numismatic gold coins were exempt.

Americans were ordered to turn their gold over to the government in exchange for a below-market price of just over $20 per ounce. Once most had complied, FDR raised gold's official price in two steps to $35 per ounce, with government pocketing this profit.

Until FDR's confiscation, American currency had been on a Gold Standard. Corporate contracts often included "gold clauses" specifying that obligations were to be paid in either dollars or a specific quantity of gold, whichever was more valuable at the time of paying off the agreement. FDR's prohibition nullified gold clauses by making possession of gold bullion illegal.

In America's Great Debasement, however, the U.S. Government allowed citizens to keep and use their silver – and continued to issue paper Silver Certificate currency. Redemption of these certificates for actual silver ended in mid-1968.

In the aftermath of its Great Debasement, writes Pamuk, the Ottoman government adopted a measure "to prohibit the use and sale of gold and silver in local markets and ordered that these be surrendered to the imperial mint at below-market prices....the state also obtained revenue from the old coins brought to the mint by the public."

Here, too, the parallels between the American and Ottoman Great Debasements are striking.

"A Plunder Machine"

The Ottoman Turks were first mentioned in Western history around 1227 A.D., when this nomadic people began settling in Turkey. More innovative and better fighters than most neighboring tribes, this people named for their first great leader Osman quickly expanded the territory under their rule.

In 1453 A.D. the Ottomans captured Constantinople, capital of the Byzantine Empire. The Muslim Ottomans were soon replacing this defeated empire with their own, ruling lands across North Africa and the Middle East. The Ottoman Empire would endure for more than 600 years.

"The Ottoman state was a plunder machine," wrote economic historian Eric L. Jones in his 1981 work *The European Miracle: Environments, Economies and Geopolitics in the History of Europe and Asia*, "which needed booty or land to fuel itself, to pay its way, to reward its officer class.."

"The Ottomans had filled a power vacuum – had taken over a region once strong, now enfeebled, looting as they went," wrote David Landes in his 1998 study *The Wealth and Poverty of Nations: Why Some Are So Rich and Some So Poor*.

They turned to the Great Debasement, we infer from Landes, because as Europe grew stronger the Ottomans "could no longer take from outside. They had to generate wealth from within, to promote productive development."

An Odd Empire

"Instead, they resorted to habit and tried to pillage the interior, to squeeze their own subjects," wrote Landes. "Nothing, not even the wealth of high officials, was secure. Nothing could be more destructive. The only thing that saved the empire from disintegration was its inefficiency, the venality of its officials, and the protective interests of stronger powers" such as Great Britain.

This was an odd empire that granted Samuel Morse his first patent for the telegraph and used his technology, yet like some more dogmatic Muslim nations it refused the printing press as "a potential instrument of sacrilege and heresy," wrote Landes.

After the worst of the Great Debasement, in 1844 the Ottoman Empire adapted monetary reforms. "The old coinage was withdrawn," wrote Lord Kinross in his 1977 history *The Ottoman Centuries: The Rise and Fall of the Turkish Empire*, "and a new one introduced on European lines, based on a gold pound."

"Fourteen Tons of Gold Leaf"

In 1843, however, the Sultan began construction of the new Dolmabahce Palace, which was designed to impress the West. According to Harvard economic historian Niall Ferguson, "Fourteen tons of gold leaf were used to gild the palace ceilings."

Centuries earlier, the Ottomans had grown wealthy as middlemen selling Asian silk and other products to Europe, and it had used that wealth to conquer Greece and several Balkan nations, and to send armies unsuccessfully to threaten the gates of Vienna.

By the 1800s Ottoman backwardness had turned them into net importers, dependent on Europe not only for merchandise, but also for loans to buy that merchandise. They used borrowed money as well to wage the Crimean and other wars against Russian incursion, and to fight uprisings in Greece (which won its independence in 1829) and elsewhere in the Empire.

Here, too, the parallels to America's Great Debasement are telling. We do not think of ourselves as an Empire, but we pay a high price to maintain troops and to fight in many nations as if we were an Empire.

And like the Ottomans, we increasingly fund our government, our military activities, our costly social programs, and even our foreign aid to other nations using borrowed money.

This became the Ottoman Empire's final undoing. European creditors used the Empire's debt as leverage to occupy Egypt.

The Sick Man of Europe

In World War I the Ottoman Empire sided with Germany and the Axis powers, the losing side. Even before this, Turkey was called "the sick man of Europe," made all the sicker by its wartime expenses, debt and casualties.

One Turkish hero who emerged during World War I was Mustafa Kemal, later known as Ataturk, "the father of the Turks," who in 1924 with his "Young Turks" toppled the Ottoman Empire in a coup.

Ataturk believed that Turkey had to westernize as rapidly as possible. He changed the nation's alphabet from Arabic letters to Roman. He prohibited women wearing the veil, and men wearing the Fez. He granted women the right to vote in Turkey before women could vote in Great Britain. He advocated a secular, democratic state oriented westward.

Turkey's democracy lately has chosen leaders who are less secular and more Islamic.

Ottoman Reflections

What can we learn from the disquieting similarities between the Ottoman Great Debasement from 1808 until 1844 and America's Great Debasement that began in 1913?

America's Founders bequeathed to us a constitutional republic. If we had honored the constraints on government in our Constitution, we likely would have escaped having a Great Debasement.

Progressives have relentlessly undermined or circumvented the chains put into the Constitution to limit the government.

Progressives amended the Constitution to permit income taxation, which the Framers carefully prohibited.

Progressives created the Federal Reserve to move the United States from solid, gold-backed currency to today's unbacked "elastic" paper fiat currency that can be printed without limit.

Progressives, by outlawing gold ownership in 1933 and then in 1971 severing the last anchor our dollar had to the stability of gold, have set us on the road to debt and debasement that led past empires – from Athens to Rome to the Ottomans to the British Empire – to collapse.

How far from a currency collapse are we? Our dollar has ticked down to its last two cents of the purchasing power it had when the Fed and income tax began in 1913. Two ticks remain on the dollar doomsday clock.

A Black Hole of Debt

The Ottoman Empire began to slide into irredeemable debasement in 1808, yet caught themselves 36 years later, in 1844, just before crossing the "event horizon" of a Black Hole of debt from which escape would be impossible.

Insatiable spending on things like gold leaf dragged the Ottoman Empire into unpayable debts and a financial death spiral by 1888, and 36 years later Ataturk toppled a hollow empire whose core had simply rotted away.

Is it only a coincidence that America began its current collapse into the Great Recession in late 2007, exactly 36 years after President Richard Nixon severed the last anchor that tied the U.S. Dollar to gold?

By 1804, the Ottoman Great Debasement had already spread from its currency to the personal values and morals of the people themselves.

"As a result of...intellectual segregation, technical lag, and industrial dependency, the balance of economic forces tilted steadily against the Ottomans," wrote Landes, "while a series of military defeats undermined their assumptions of superiority and paralyzed their ability to respond...."

"A byzantine bureaucracy made everything harder with thorny regulations in incomprehensible officialese," Landes continued. "Corruption – the only way to get something done – just fed on itself."

The Ottoman bureaucracy was originally based on individual merit, but, as Niall Ferguson wrote, this gradually deteriorated so that "selection and promotion seemed to depend more on bribery and favouritism than on aptitude; the rate of churn became absurdly high as people jostled for the perquisites of office," while administrative standards deteriorated.

By 1807, wrote Ferguson, "the Ottoman army appeared to be run primarily for the enrichment and convenience of its officers."

Lord Kinross describes one of the most ardent young reformers of this period "weakening in moral conviction and declining in spirit.... He fell into debt and became prone to corruption. The Sultan himself, wearying of reform, became increasingly irresolute in public affairs...."

Undoing Debasement

Look deeply into the Ottoman mirror. Does this seem familiar? The Great Debasement not only of money but of the core values that keep a society healthy and strong? Paralyzing military morasses from Vietnam to Afghanistan? Crony capitalism and political corruption at the highest levels? Bureaucrats and politicians who hamstring the honest with regulations and ridiculous rules of engagement, and who channel rewards to the dishonest through cronyism and wealth redistribution? And above all, an eagerness to "take the cash and let the credit go," to spend recklessly and run up debt in the expectation that inflation or Uncle Sugar will cover the tab?

Bastiat was right. As America turns into a European welfare state, our government has become the great fiction by which everyone tries to live at the expense of everyone else.

We need to tell the Progressives who believe in no God (except the state) that if there is no God, then there also is no Santa Claus.

Have you ever met a Progressive who did not believe that the government is Santa Claus with a bottomless bag of free goodies?

In August 2012 a new Santa revealed himself.

"The ultimate purpose of economics, of course, is to understand and promote the enhancement of well-being," said Federal Reserve Chairman Ben Bernanke.

Bernanke gushed enthusiastically about the Himalayan kingdom of Bhutan, "which abandoned tracking gross national product in 1972 in favor of its Gross National Happiness index."

"We should seek better and more-direct measurements of economic well-being, the ultimate objective of our policy decisions," said Mr. Bernanke, who mentioned that two years earlier he had given a speech titled "The Economics of Happiness."

The "Happiness State"

Richard Ebeling, a senior fellow at the American Institute for Economic Research, was not overjoyed by the Fed Chairman's new self-appointed mandate.

"The happiness advocate," he wrote, "concludes that there is only one cure....to use the power of the state to socially engineer better and happier outcomes for the vast majority of mankind....through government 'reeducation.' Schools and other educational vehicles should be applied by the state to inform and teach people proper values and virtues to make them better and happier."

A Bernanke-engineered "Happiness State," notes Ebeling, would not be based on individual values or on small groups and organizations. It would be a one-size-fits-all nanny state to make New York City Mayor Michael Bloomberg proud.

The Declaration of Independence identifies "unalienable rights, that among these are Life, Liberty and the Pursuit of Happiness" and that the People shall establish "new government laying its Foundation on such Principles... as to them shall seem most likely to effect their Safety and Happiness."

As we read this, it seems to affirm that such decisions should be left to individuals, not the Progressive Collective, which has undermined the Foundation our Founders established.

The answer to America's Great Debasement, not only of our money's value but of our values as well, is to return to what is solid – in the kind of reliable money they intended, and in the original Foundation they set firmly on the cornerstones of individual responsibility, self-reliance and small government.

The remedy for the Great Debasement is a Great Re-basement, a restoration of the base, the Foundation, on which our Republic's greatness was built.

"You have to choose between trusting
to the natural stability of gold and
the natural stability of the honesty and intelligence
of the members of the Government.

"And, with due respect for these gentlemen,
I advise you, as long as the Capitalist system lasts,
to vote for gold."

– George Bernard Shaw
Irish Playwright

Chapter Notes

Chapter One:
A View from the Cliff

Curtis Dubay, "Ernst & Young: Obama's Tax Increase Would Kill 710,000 Jobs," Heritage Foundation, July 18, 2012. URL: http://blog.heritage.org/2012/07/18/ernst-and-young-obamas-tax-increase-would-kill-710000-jobs/print/

Drs. Robert Carroll and Gerald Prante, "Long-run Macroeconomic Impact of Increasing Tax Rates on High-income Taxpayers in 2013." (Monograph). Ernst & Young, July 2012. URL: http://majorityleader.gov/uploadedfiles/Ernst_And_Young_Study_July_2012.pdf

Joel Gehrke, "55 Percent of Small Business Owners Would Not Start Company Today, Blame Obama," *The Washington Examiner*, September 26, 2012.

Christopher J. Conover, "Congress Should Account for the Excess Burden of Taxation," Cato Institute, Policy Analysis # 669, October 13, 2010. URL: http://www.cato.org/pubs/pas/PA669.pdf

Richard Duncan, *The Dollar Crisis: Causes, Consequences, Cures*. Singapore: John Wiley & Sons (Asia), 2003.

____________, *The New Depression: The Breakdown of the Paper Money Economy.* New York: John Wiley & Sons, 2012.

Peter Ferrara, *America's Ticking Bankruptcy Bomb: How the Looming Debt Crisis Threatens the American Dream – and How We Can Turn the Tide Before It's Too Late.* New York: Broadside Books, 2011.

Laurence J. Kotlikoff and Scott Burns, *The Coming Generational Storm: What You Need to Know About America's Economic Future.* Cambridge, Massachusetts: MIT Press, 2005.

Stephen Moore, "2016: A Fractured America" (Cover Story), *Newsmax* Magazine, October 2012.

Charles Murray, *Coming Apart: The State of White America, 1960-2010*. New York: Crown Forum, 2012.

Johan Norberg, *Financial Fiasco: How America's Infatuation with Home Ownership and Easy Money Created the Economic Crisis*. Washington, D.C.: Cato Institute, 2009.

Grover Norquist and John R. Lott, Jr., *Debacle: Obama's War on Jobs and Growth and what We Can do Now to Regain Our Future.* New York: Wiley, 2012.

Carmen M. Reinhart and Kenneth S. Rogoff, *This Time Is Different: Eight Centuries of Financial Folly*. Princeton, New Jersey: Princeton University Press, 2009.

Craig R. Smith and Lowell Ponte, *Crashing the Dollar: How to Survive a Global Currency Collapse.* Phoenix: Idea Factory Press, 2010.

________________________, *The Inflation Deception: Six Ways Government Tricks Us...And Seven Ways to Stop It!* Phoenix: Idea Factory Press, 2011.

Citizen's Guide to the 2011 Financial Report of the United States Government, Washington, D.C.: U.S. Department of the Treasury, 2012. URL: http://www.fms.treas.gov/fr/11frusg/11frusg.pdf

Emily Goff, Romina Boccia and John Fleming, *2012 Federal Budget in Pictures: Budget Chart Book.* Washington, D.C.: Heritage Foundation/Thomas A. Roe Institute for Economic Policy Studies, 2012.

Chapter Two:
Ascent of the Dollar

Milton Friedman & Anna Jacobson Schwartz, *A Monetary History of the United States, 1867-1960.* A Study by the National Bureau of Economic Research, New York. Princeton, New Jersey: Princeton University Press, 1963.

Craig Karmin, *Biography of the Dollar: How the Mighty Buck Conquered the World and Why It's Under Siege.* New York: Crown Business, 2008.

Murray N. Rothbard, *A History of Money and Banking in the United States: The Colonial Era to World War II.* Auburn, Alabama: Ludwig von Mises Institute, 2002. This can be downloaded from the Internet at no cost from http://mises.org/Books/HistoryofMoney.pdf

________________, *What Has Government Done to Our Money?* Auburn, Alabama: Ludwig von Mises Institute, 2008. This can be downloaded from the Internet at no cost from http://mises.org/Books/Whathasgovernmentdone.pdf

Gerald P. Dwyer, Jr., "Wildcat Banking, Banking Panics, and Free Banking in the United States." *Economic Review* / Federal Reserve bank of Atlanta, December 1996. URL: Http://www.frbatlanta.org/filelegacydocs/acfce.pdf

Chapter Three:
The Progressive Takeover

Richard A. Epstein, *How Progressives Rewrote the Constitution.* Washington, D.C.: Cato Institute, 2006.

Charles R. Kesler, *I Am the Change: Barack Obama and the Crisis of Liberalism.* New York: Broadside Books, 2012.

Chapter Four:
The Great Debasement

Liaquat Ahamed, *Lords of Finance: The Bankers Who Broke the World.* New York: Penguin Books, 2009.

John H. Cochrane, "The Future of Central Banks," *Wall Street Journal*, August 31, 2012. URL: http://johnhcochrane.blogspot.com/2012/08/the-future-of-central-banks.html

Barry Eichengreen, *Exorbitant Privilege: The Rise and Fall of the Dollar and the Future of the International Monetary System.* Oxford: Oxford University Press, 2011.

_______________, *Global Imbalances and the Lessons of Bretton Woods* (Cairoli Lectures). Cambridge, Massachusetts: MIT Press, 2010.

_______________, *Globalizing Capital: A History of the International Monetary System* (Second

Edition).

______________, *Golden Fetters: The Gold Standard and the Great Depression, 1919-1939* (NBER Series on Long-Term Factors in Economic Development). Oxford: Oxford University Press, 1996.

Barry Eichengreen and Marc Flandreau, *Gold Standard In Theory & History.* London: Routledge, 1997.

Ralph T. Foster, *Fiat Paper Money: The History and Evolution of Our Currency.* Second Edition. Monograph, 2010.

Milton Friedman & Anna Jacobson Schwartz, *A Monetary History of the United States, 1867-1960.* A Study by the National Bureau of Economic Research, New York. Princeton, New Jersey: Princeton University Press, 1963.

J.D. Gould, *The Great Debasement: Currency and the Economy in Mid-Tudor England.* London: Clarendon, 1970.

William Greider, *Secrets of the Temple: How the Federal Reserve Runs the Country.* New York: Simon & Schuster, 1989.

G. Edward Griffin, *The Creature from Jekyll Island: A Second Look at the Federal Reserve.* Third Edition. Westlake Village, California: American Media, 1998.

Robert L. Hetzel, *The Great Recession: Market Failure or Policy Failure? (Studies in Macroeconomic History).* New York: Cambridge University Press, 2012.

______________, *The Monetary Policy of the Federal Reserve.* New York: Cambridge University Press, 2008.

Arnold Kling, *The Case for Auditing the Fed Is Obvious.* (Monograph / Briefing Paper). Washington, D.C.: Cato Institute, April 27, 2010. URL: http://www.cato.org/pubs/bp/bp118.pdf

Martin Mayer, *The Fed: The Inside Story of How the World's Most Powerful Financial Institution Drives the Markets.* New York: Free Press, 2001.

John D. Munro, *The Coinages and Monetary Policies of Henry VIII (r. 1509-1547): Contrasts Between Defensive and Aggressive Debasements* (Monograph). Toronto: University of Toronto Economics Department, 2010. Access via URL: http://ideas.repec.org/p/tor/tecipa/tecipa-417.html

Maxwell Newton, *The Fed: Inside the Federal Reserve, the Secret Power Center that Controls the American Economy.* New York: Times Books, 1983.

Ron Paul, *End The Fed.* New York: Grand Central Publishing / Hachette, 2009.

Arthur J. Rolnick, Francois R. Velde and Warren E. Weber, "The Debasement Puzzle: An Essay on Medieval Monetary History," *Federal Reserve Bank of Minneapolis Quarterly Review*, Vol. 21 #4 (Fall 1997). URL: http://www.minneapolisfed.org/research/qr/qr2142.ps

Murray N. Rothbard, *The Case Against the Fed.* Second Edition. Auburn, Alabama: Ludwig von Mises Institute, 2007. A version of this book can be downloaded from the Internet at no cost from http://mises.org/books/Fed.pdf

________________, *A History of Money and Banking in the United States: The Colonial Era to World War II.* Auburn, Alabama: Ludwig von Mises Institute, 2002. This can be downloaded from the Internet at no cost from http://mises.org/Books/HistoryofMoney.pdf

________________, *The Mystery of Banking*. Second Edition. Auburn, Alabama: Ludwig von Mises Institute, 2008. This can be downloaded from the Internet at no cost from http://mises.org/Books/MysteryofBanking.pdf

________________, *What Has Government Done to Our Money?* Auburn, Alabama: Ludwig von Mises Institute, 2008. This can be downloaded from the Internet at no cost from http://mises.org/Books/Whathasgovernmentdone.pdf

Detlev S. Schlichter, *Paper Money Collapse: The Folly of Elastic Money and the Coming Monetary Breakdown.* New York: John Wiley & Sons, 2011.

Yi Wen, "Monetary Policy's Effects on Unemployment," Federal Reserve Bank of St. Louis *Economic Synopses*, # 10 (2011). URL: http://research.stlouisfed.org/publications/es/11/ES1110.pdf

Text of H.R. 7837, legislation signed into law as the Federal Reserve Act by President Woodrow Wilson, December 23, 1913. URL: http://www.llsdc.org/attachments/files/105/FRA-LH-PL63-43.pdf

Chapter Five:
The Age of "Modern Money"

About Modern Monetary Theory:

L. Randall Wray, *Modern Money Theory: A Primer on Macroeconomics for Sovereign Monetary Systems.* London: Palgrave Macmillan, 2012.

____________, *Understanding Modern Money: The Key to Full Employment and Price Stability.* Northampton, Massachusetts: Edward Elgar Publishing, 2006.

Dr. Wray's online primer about Modern Monetary Theory can be read at URL: http://neweconomicperspectives.org/p/modern-monetary-theory-primer.html

Dylan Matthews, "Modern Monetary Theory: An Unconventional Take on Economic Strategy," *Washington Post*, February 18, 2012. URL: http://www.washingtonpost.com/business/modern-monetary-theory-is-an-unconventional-take-on-economic-strategy/2012/02/15/gIQAR8uPMR_story.html

Winston Gee, "Debts, Deficits, and Modern Monetary Theory" (interview with Bill Mitchell). *Harvard International Review*, October 16, 2011. URL: http://hir.harvard.edu/debt-deficits-and-modern-monetary-theory

Bill Mitchell, "There is No Financial Crisis So Deep That It Cannot Be Dealt With by Public Spending," October 11, 2010. URL: http://bilbo.economicoutlook.net/blog/?p=11854

Warren Mosler, "Modern Monetary Theory: The Last Progressive Left Standing," *Huffington Post*, June 13, 2011. URL: http://www.huffingtonpost.com/warren-mosler/modern-monetary-theory-th_b_872449.html

John Carney, "Modern Monetary Theory and Austrian Economics," CNBC, December 27, 2011. URL: http://www.cnbc.com/id/45795986/Modern_Monetary_Theory_and_Austrian_Economics

__________, "Would Modern Monetary Theory Lead to Crony Capitalism?" CNBC, August 15, 2011.

__________, "Modern Monetary Theory, Crony Capitalism and the Tea Party," CNBC, December

22, 2011.

__________, "Government Spending and Monetary Theory," CNBC, August 11, 2011.

About Financial Repression:

Carmen M. Reinhart, "Financial Repression Back to Stay," *Bloomberg*, March 11, 2012. URL: http://www.bloomberg.com/news/2012-03-11/financial-repression-has-come-back-to-stay-carmen-m-reinhart.html

Carmen M. Reinhart and M. Belen Sbrancia, "The Liquidation of Government Debt," National Bureau of Economic Research (NBER) Working Paper # 16893. March 2011. URL: http://www.imf.org/external/np/seminars/eng/2011/res2/pdf/crbs.pdf

Alberto Giovanni and Martha De Melo, "Government Revenues from Financial Repression," American Economic Review, Vol. 83 #4 (September 1993). URL: http://www.jstor.org/discover/10.2307/2117587?uid=3739560&uid=2&uid=4&uid=3739256&sid=21101221127691

Buttonwood, "Carmen Reinhart and Financial Repression," *The Economist*, January 10, 2012. URL: http://www.economist.com/blogs/buttonwood/2012/01/debt-crisis/print

Member of the European Parliament Nigel Farage, "Europe Is About to Impose Extreme Repression," *King World News* (Interview), June 22, 2012. URL: http://kingworldnews.com/kingworldnews/KWN_DailyWeb/Entries/2012/6/22_Nigel_Farage_-_Europe_is_About_to_Impose_Extreme_Repression.html

Chapter Six:
The End of the Republic

Charles R. Kesler, *I Am the Change: Barack Obama and the Crisis of Liberalism.* New York: Broadside Books, 2012.

Bob Woodward, *The Power of Politics.* New York: Simon & Schuster, 2012.

British correspondent reference to President Franklin Delano Roosevelt's rule as America's "fling at National Socialism" appears in Alistair Cooke, *Alistair Cooke's America.* New York: Alfred A. Knopf, 1982, pages 329 and 331.

Chapter Seven:
The Mandate of Heaven

Lowell Ponte, "The Great Blizzard of '88," *Reader's Digest*, December 1987.

Lowell Ponte, *The Cooling*. Englewood Cliffs, New Jersey: Prentice-Hall, 1976.

__________, "Food: America's Secret Weapon," *Reader's Digest*, May 1982.

__________, "What's Wrong With Our Weather," *Reader's Digest*, November 1988.

__________, "Who Will Control the Weather?" *Reader's Digest*, May 1980.

__________, "Why All the Crazy Weather?" *Reader's Digest*, December 1993.

___________, "Why Our Weather is Going Wild," *Reader's Digest*, December 1982.

___________, "Why They Still Can't Predict the Weather," *Reader's Digest*, September 1992.

Lowell Ponte and Mark Morano, "Al Gore Spins Global Warming" (Cover Story), *Newsmax Magazine*, July 2006.

Kevin McCoy, "Analysis: Nation's Water Costs Rushing Higher," *USA Today*, September 28, 2012. URL: http://www.cnbc.com/id/49211886

Matthew L. Wald, "A Fine for Not Using a Biofuel That Doesn't Exist," *New York Times*, January 9, 2012. URL: http://www.nytimes.com/2012/01/10/business/energy-environment/companies-face-fines-for-not-using-unavailable-biofuel.html

Chapter Eight:
The Dollar Engulfed

Kevin D. Freeman, *Economic Warfare: Risks and Responses: Analysis of Twenty-First Century Risks in Light of the Recent Market Collapse* (Monograph). Cross Consulting and Services, 2009. This can be downloaded from the Internet at no cost from

http://av.r.ftdata.co.uk/files/2011/03/49755779-Economic-Warfare-Risks-and-Responses-by-Kevin-D-Freeman.pdf or at no cost from

http://www.freemanglobal.com/uploads/Economic_Warfare_Risks_and_Responses.pdf

On Israel offshore natural gas, see Ethan Bronner, "Gas Field Confirmed Off Coast of Israel," New York Times, December 30, 2010. URL: http://www.nytimes.com/2010/12/31/world/middleeast/31leviathan.html

On Irish offshore oil, see Mark P. Mills, "America Take Note: Technology Unleashes Black Gold to Rescue Ireland's Economy," Forbes Magazine, September 18, 2012. URL: http://www.forbes.com/sites/markpmills/2012/09/18/america-take-note-technology-unleashes-black-gold-to-rescue-irelands-economy/

Chapter Nine:
The Dollar and Its Enemies

Barry Eichengreen, *Exorbitant Privilege: The Rise and Fall of the Dollar and the Future of the International Monetary System.* Oxford: Oxford University Press, 2011.

_______________, *Global Imbalances and the Lessons of Bretton Woods* (Cairoli Lectures). Cambridge, Massachusetts: MIT Press, 2010.

_______________, "Why the Dollar's Reign Is Near an End," *Wall Street Journal*, March 1, 2011. URL: http://online.wsj.com/article/SB10001424052748703313304576132170181013248.html

Henry Hazlitt, *The Failure of the "New Economics": An Analysis of The Keynesian Fallacies.* New Rochelle, New York: Arlington House, 1959.

___________, *From Bretton Woods to World Inflation: A Study of Causes & Consequences.* Chicago: Regnery Gateway, 1984. This can be downloaded from the Internet at no cost from http://mises.org/books/brettonwoods.pdf

James Rickards, *Currency Wars: The Making of the Next Global Crisis*. New York: Portfolio/Penguin, 2011. See pages 3-34.

Chapter Ten:
Ghost Money: After the Dollar's Nine Lives

Joel Kurtzman, *The Death of Money: How the Electronic Economy Has Destabilized the World's Markets and Created Financial Chaos.* New York: Simon & Schuster, 1993.

"The Last Days of Cash: How E-Money Technology is Plugging Us into the Digital Economy (Special Report, includes many articles), *IEEE Spectrum*, May 30, 2012. URL: http://spectrum.ieee.org/static/future-of-money

Chapter Eleven:
New Dark Age or Golden Age?

John Butler, *The Golden Revolution: How to Prepare for the Coming Global Gold Standard.* New York: John Wiley & Sons, 2012.

Harold Van B. Cleveland and others, *Money and the Coming World Order*, Second Edition. Greenwich, Connecticut: Lehrman Institute, 2012.

Lewis E. Lehrman, *The True Gold Standard – A Monetary Reform Plan Without Official Reserve Currencies*. Greenwich, Connecticut: The Lehrman Institute, 2011.

Nathan Lewis and Addison Wiggin, *Gold: The Once and Future Money.* Hoboken, New Jersey: John Wiley & Sons, 2007.

Ron Paul and Lewis Lehrman, *The Case for Gold: A Minority Report of the U.S. Gold Commission.* Auburn, Alabama: Ludwig von Mises Institute, 2007. This can be downloaded from the Internet at no cost from http://mises.org/books/caseforgold.pdf

Ayn Rand, *Capitalism: The Unknown Ideal (With additional articles by Nathaniel Branden, Alan Greenspan, and Robert Hessen).* New York: Signet / New American Library, 1967.

Craig R. Smith, "The Islamic Gold Dinar Conspiracy?" *Israel Forum*, July 15, 2005. URL: http://www.israelforum.com/board/archive/index.php/t-8416.html

Epilogue:
The Ottoman Mirror

Sevket Pamuk, "The Great Ottoman Debasement, 1808-1844: A Political Economy Framework," in The Ataturk Institute, New Dimensions of Modernizing Processes. URL: http://www.ata.boun.edu.tr/faculty/Faculty/Sevket%20Pamuk/publications/Pamuk,%20Sevket_The%20Great%20Ottoman%20Debasement%201808%201844.pdf

___________, *A Monetary History of the Ottoman Empire*. New York: Cambridge University Press, 2000.

Lord Kinross, *The Ottoman Centuries: The Rise and Fall of the Turkish Empire*. New York: Morrow, 1977. pp. 479-481.

Niall Ferguson, *Civilization: The West and the Rest*. New York: Penguin, 2011. pp. 71-72, 87-88.

Eric L. Jones, *The European Miracle: Environments, Economies and Geopolitics in the History of Europe and Asia*. Cambridge, England: Cambridge University Press, 1981.

David S. Landes, *The Wealth and Poverty of Nations: Why Some Are So Rich and Some So Poor*. New York: W.W. Norton, 1998. Page 402.

Ben S. Bernanke, "Economic Measurement" (speech transcript), August 6, 2012. Board of Governors of the Federal Reserve System. URL: http://www.federalreserve.gov/newsevents/speech/bernanke20120806a.htm

Richard M. Ebeling, "The New 'Happiness' Economics: An Austrian Critique" (Monograph), August 2012. URL: http://mises.org/journals/scholar/ebeling2.pdf

Sources

Liaquat Ahamed, *Lords of Finance: The Bankers Who Broke the World.* New York: Penguin Books, 2009.

George A. Akerlof and Robert J. Schiller, *Animal Spirits: How Human Psychology Drives the Economy, and Why It Matters for Global Capitalism.*
Princeton, New Jersey: Princeton University Press, 2009.

John Anthers, *The Fearful Rise of Markets: Global Bubbles, Synchronized Meltdowns, and How to Prevent Them In The Future.* London: FT Press, 2010.

William W. Beach and others, *Obama Tax Hikes: The Economic and Fiscal Effects* (Monograph). Washington, D.C.: Heritage Foundation, 2010.

David Beckworth (Editor), *Boom and Bust Banking: The Causes and Cures of the Great Recession.* Oakland, California: Independent Institute, 2012.

Ben S. Bernanke and others, *Inflation Targeting: Lessons from the International Experience.* Princeton, New Jersey: Princeton University Press, 1999.

Peter Bernholz, *Monetary Regimes and Inflation: History, Economic and Political Relationships.* Williston, Vermont: Edward Elgar Publishing, 2006.

William Bonner and Addison Wiggin, *Financial Reckoning Day: Surviving the Soft Depression of the 21st Century.* Hoboken, New Jersey: John Wiley & Sons, 2004.

________________________________, *The New Empire of Debt: The Rise and Fall of an Epic Financial Bubble* (Second Edition). Hoboken, New Jersey: John Wiley & Sons, 2009.

Neal Boortz and John Linder, *The FairTax Book: Saying Goodbye to the Income Tax and the IRS....* New York: Regan Books / HarperCollins, 2005.

Neal Boortz, John Linder and Rob Woodall, *FairTax: The Truth: Answering the Critics*. New York: Harper, 2008.

Jerry Bowyer, *The Free Market Capitalist's Survival Guide: How to Invest and Thrive in an Era of Rampant Socialism.* New York: Broad Side / Harper Collins, 2011.

H.W. Brands, *The Age of Gold: The California Gold Rush and the New American Dream.* New York: Doubleday / Random House, 2002.

Arthur C. Brooks, *The Battle: How the Fight Between Free Enterprise and Big Government Will Shape America's Future.* New York: Basic Books / Perseus, 2010.

______________, *Gross National Happiness: Why Happiness Matters for America – and How We Can Get More of It.* New York: Basic Books, 2008.

______________, *The Road to Freedom: How to Win the Fight for Free Enterprise*. New York: Basic Books/Perseus Books, 2012.

James M. Buchanan and Richard E. Wagner, *Democracy in Deficit: The Political Legacy of Lord Keynes.* Indianapolis: Liberty Fund, 1999.

Todd G. Buchholz, *New Ideas from Dead Economists: An Introduction to Modern Economic Thought*. New York: New American Library/Penguin Books, 1989.

John Butler, *The Golden Revolution: How to Prepare for the Coming Global Gold Standard.* New York: John Wiley & Sons, 2012.

Bruce Caldwell (Editor), *The Collected Works of F.A. Hayek, Volume 2: The Road to Serfdom: Texts and Documents: The Definitive Edition.* Chicago: University of Chicago Press, 2007.

Stephen G. Cecchetti and others, *The Real Effects of Debt*. BIS Working Papers No 352. Basel, Switzerland: Bank for International Settlements, September 2011. URL: http://www.bis.org/publ/work352.pdf

Marc Chandler, *Making Sense of the Dollar: Exposing Dangerous Myths about Trade and Foreign Exchange.* New York: Bloomberg Press, 2009.

Harold Van B. Cleveland and others, *Money and the Coming World Order*, Second Edition. Greenwich, Connecticut: Lehrman Institute, 2012.

Congressional Budget Office, *The Budget and Economic Outlook: Fiscal Years 2011 to 2021.* Washington, D.C.: Congressional Budget Office, January 2011. URL: http://www.cbo.gov/ftpdocs/120xx/doc12039/01-26_FY2011Outlook.pdf

Jerome R. Corsi, *America for Sale: Fighting the New World Order, Surviving a Global Depression, and Preserving U.S.A. Sovereignty.* New York: Threshold Editions / Simon & Schuster, 2009.

Crews, Clyde Wayne, Jr., *Ten Thousand Commandments: An Annual Snapshot of the Federal Regulatory State.* 2011 Edition. (Monograph). Washington, D.C.: Competitive Enterprise Institute, 2011.

Glyn Davies, *A History of Money: From Ancient Times to the Present Day*. Third Edition. Cardiff: University of Wales Press, 2002.

Glyn Davies and Roy Davies, *A Comparative Chronology of Money: Monetary History from Ancient Times to the Present Day.* (Monograph based on Glyn Davies and Roy Davies, above.) (2006) URL: http://projects.exeter.ac.uk/RDavies/arian/amser/chrono.html

Hernando de Soto, *The Mystery of Capital: Why Capitalism Triumphs in the West and Fails Everywhere Else*. New York: Basic Books / Perseus, 2000.

Peter H. Diamandis and Steven Kotler, *Abundance: The Future Is Better Than You Think.* New York: Free Press, 2012.

Jared Diamond, *Collapse: How Societies Choose to Fail or Succeed.* New York: Viking Press, 2005.

Peter F. Drucker, *Post-Capitalist Society.* New York: Harper Business, 1993.

Dinesh D'Souza, *Obama's America: Unmaking the American Dream*. Washington, D.C.: Regnery, 2012.

_____________, *The Roots of Obama's Rage*. Washington, D.C.: Regnery, 2010.

_____________, *The Virtue of Prosperity: Finding Values in an Age of Techno-Affluence.* New York: Free Press / Simon & Schuster, 2000.

Richard Duncan, *The Dollar Crisis: Causes, Consequences, Cures*. Singapore: John Wiley & Sons (Asia), 2003.

_____________, *The New Depression: The Breakdown of the Paper Money Economy.* New York: John Wiley & Sons, 2012.

Gregg Easterbrook, *The Progress Paradox: How Life Gets Better While People Feel Worse*. New York: Random House, 2003.

Gauti B. Eggertsson, *What Fiscal Policy Is Effective at Zero Interest Rate?* Staff Report No. 402 (Monograph). New York: Federal Reserve Bank of New York, November 2009. URL: http://www.newyorkfed.org/research/staff_reports/sr402.pdf

Barry Eichengreen, *Exorbitant Privilege: The Rise and Fall of the Dollar and the Future of the International Monetary System.* Oxford: Oxford University Press, 2011.

_______________, *Global Imbalances and the Lessons of Bretton Woods* (Cairoli Lectures). Cambridge, Massachusetts: MIT Press, 2010.

_______________, *Globalizing Capital: A History of the International Monetary System* (Second Edition).

_______________, *Golden Fetters: The Gold Standard and the Great Depression, 1919-1939* (NBER Series on Long-Term Factors in Economic Development). Oxford: Oxford University Press, 1996.

Barry Eichengreen and Marc Flandreau, *Gold Standard In Theory & History.* London: Routledge, 1997.

Richard A. Epstein, *How Progressives Rewrote the Constitution*. Washington, D.C.: Cato Institute, 2006.

_______________, *Takings: Private Property and the Power of Eminent* Domain. Cambridge, Massachusetts: Harvard University Press, 1985.

Niall Ferguson, *The Ascent of Money: A Financial History of the World*. New York: Penguin Press, 2008.

____________, *The Cash Nexus: Money and Power in the Modern World, 1700-2000*. New York: Basic Books, 2002.

____________, *Civilization: The West and the Rest*. New York: Penguin Books, 2011.

____________, *Colossus: The Price of America's Empire*. New York: Penguin Press, 2004.

Peter Ferrara, *America's Ticking Bankruptcy Bomb: How the Looming Debt Crisis Threatens the American Dream – and How We Can Turn the Tide Before It's Too Late*. New York: Broadside Books, 2011.

Ralph T. Foster, *Fiat Paper Money: The History and Evolution of Our Currency*. Second Edition. 2008.

Justin Fox, *The Myth of the Rational Market: A History of Risk, Reward, and Delusion on Wall Street*. New York: Harper Business, 2009.

Kevin D. Freeman, *Economic Warfare: Risks and Responses: Analysis of Twenty-First Century Risks in Light of the Recent Market Collapse* (Monograph). Cross Consulting and Services, 2009. This can be downloaded from the Internet at no cost from http://av.r.ftdata.co.uk/files/2011/03/49755779-Economic-Warfare-Risks-and-Responses-by-Kevin-D-Freeman.pdf

or at no cost from
http://www.freemanglobal.com/uploads/Economic_Warfare_Risks_and_Responses.pdf

George Friedman, *The Next Decade: Where We've Been...And Where We're Going*. New York: Doubleday, 2011.

Milton Friedman, *An Economist's Protest.* Second Edition. Glen Ridge, New Jersey: Thomas Horton and Daughters, 1975. Also published as *There's No Such Thing As A Free Lunch.* La Salle, Illinois: Open Court Publishing, 1975.

______________, *Capitalism & Freedom: A Leading Economist's View of the Proper Role of Competitive Capitalism.* Chicago: University of Chicago Press, 1962.

______________, *Dollars and Deficits: Inflation, Monetary Policy and the Balance of Payments.* Englewood Cliffs, New Jersey: Prentice-Hall, 1968.

______________, *Money Mischief: Episodes in Monetary History.* New York: Harcourt Brace, 1992.

______________, *On Economics: Selected Papers*. Chicago: University of Chicago Press, 2007.

Milton & Rose Friedman, *Free to Choose: A Personal Statement*. New York: Harcourt Brace Jovanovich, 1980.

____________________, *Tyranny of the Status Quo.* San Diego, California: Harcourt Brace Jovanovich, 1984.

Milton Friedman & Anna Jacobson Schwartz, *A Monetary History of the United States, 1867-1960.* A Study by the National Bureau of Economic Research, New York. Princeton, New Jersey: Princeton University Press, 1963.

Francis Fukuyama, *The End of History and The Last Man*. New York: Free Press, 1992.

_______________, *Trust: The Social Virtues and The Creation of Prosperity.* New York: Free Press, 1996.

James K. Galbraith, *The Predator State: How Conservatives Abandoned the Free Market and Why Liberals Should Too.* New York: Free Press, 2008.

John D. Gartner, *The Hypomanic Edge: The Link Between (A Little) Craziness and (A Lot of) Success in America.* New York: Simon & Schuster, 2005.

Charles Gasparino, *Bought and Paid For: The Unholy Alliance Between Barack Obama and Wall Street.* New York: Sentinel / Penguin, 2010.

Francis J. Gavin, *Gold, Dollars, and Power: The Politics of International Monetary Relations, 1958-1971 (The New Cold War History)*. Chapel Hill, North Carolina: University of North Carolina Press, 2007.

Nicole Gelinas, *After the Fall: Saving Capitalism from Wall Street – and Washington.* New York: Encounter Books, 2011.

Pamela Geller and Robert Spencer, *The Post-American Presidency.* New York: Threshold Editions / Simon & Schuster, 2010.

George Gilder, *Wealth and Poverty*. New York: Basic Books, 1981.

Jonah Goldberg, *Liberal Fascism: The Secret History of the American Left, From Mussolini to the Politics of Meaning*. New York: Doubleday, 2008.

____________, *The Tyranny of Cliches: How Liberals Cheat in the War of Ideas.* New York: Sentinel / Penguin, 2012.

Jason Goodwin, *Greenback: The Almighty Dollar and The Invention of America.* New York: John Macrae / Henry Holt and Company, 2003.

Charles Goyette, *The Dollar Meltdown: Surviving the Impending Currency Crisis with Gold, Oil, and Other Unconventional Investments.* New York: Portfolio / Penguin, 2009.

Alan Greenspan, *The Age of Turbulence: Adventures in a New World.* New York: Penguin Books, 2007.

William Greider, *Secrets of the Temple: How the Federal Reserve Runs the Country.* New York: Simon & Schuster, 1989.

G. Edward Griffin, *The Creature from Jekyll Island: A Second Look at the Federal Reserve.* Third Edition. Westlake Village, California: American Media, 1998.

Alexander Hamilton, James Madison and John Jay, *The Federalist Papers.* New York: Mentor / New American Library, 1961. Pages 280-288. For an online version of James Madison's Federalist Paper No. 44, go to this URL: http://www.constitution.org/fed/federa44.htm

Keith Hart, *Money in an Unequal World.* New York: TEXERE, 2001.

Friedrich A. Hayek (Editor), *Capitalism and the Historians.* Chicago: Phoenix Books / University of Chicago Press, 1963.

________________, *Choice in Currency: A Way to Stop Inflation.* London: Institute of Economic Affairs, 1976. This can be downloaded from the Internet at no cost from http://www.iea.org.uk/sites/default/files/publications/files/upldbook409.pdf

________________, *The Counter-Revolution of Science: Studies On The Abuse of Reason.* New York: The Free Press / Macmillan / Crowell-Collier, 1955.

________________, *The Constitution of Liberty.* The Definitive Edition, Edited by Ronald Hamowy. Chicago: University of Chicago Press, 2011.

________________, *Denationalisation of Money: The Argument Refined: An Analysis of the Theory and Practice of Concurrent Currencies. Third Edition.* London: Institute of Economic Affairs, 1990. This can be downloaded from the Internet at no cost from http://mises.org/books/denationalisation.pdf

________________, *The Fatal Conceit: The Errors of Socialism.* Chicago: University of Chicago Press, 1991.

________________, *The Road to Serfdom.* Chicago: Phoenix Books / University of Chicago Press, 1944.

Henry Hazlitt, *The Failure of the "New Economics": An Analysis of The Keynesian Fallacies.* New Rochelle, New York: Arlington House, 1959.

___________, *From Bretton Woods to World Inflation: A Study of Causes & Consequences.* Chicago: Regnery Gateway, 1984. This can be downloaded from the Internet at no cost from http://mises.org/books/brettonwoods.pdf

Robert L. Hetzel, *The Great Recession: Market Failure or Policy Failure? (Studies in Macroeco-*

nomic History). New York: Cambridge University Press, 2012.

______________, *The Monetary Policy of the Federal Reserve*. New York: Cambridge University Press, 2008.

David Horowitz and Jacob Laksin, *The New Leviathan: How the Left-Wing Money-Machine Shapes American Politics and Threatens America's Future.* New York: Crown Forum, 2012.

W.H. Hutt, *The Keynesian Episode: A Reassessment.* Indianapolis: Liberty*Press*, 1979.

Craig Karmin, *Biography of the Dollar: How the Mighty Buck Conquered the World and Why It's Under Siege.* New York: Crown Business, 2008.

Charles R. Kesler, *I Am the Change: Barack Obama and the Crisis of Liberalism.* New York: Broadside Books, 2012.

John Maynard Keynes, *Essays in Persuasion*. New York: W.W. Norton, 1963.

___________________, *The General Theory of Employment, Interest, and Money.* New York: Harcourt, Brace & World, 1935.

Arnold Kling, *The Case for Auditing the Fed Is Obvious.* (Monograph / Briefing Paper). Washington, D.C.: Cato Institute, April 27, 2010. URL: http://www.cato.org/pubs/bp/bp118.pdf

Gabriel Kolko, *Railroads and Regulation 1877-1916*. New York: W.W. Norton, 1970. Originally published in 1965 by Princeton University Press.

Laurence J. Kotlikoff, *Jimmy Stewart Is Dead: Ending the World's Ongoing Financial Plague with Limited Purpose Banking.* Hoboken, New Jersey: John Wiley & Sons, 2011.

Laurence J. Kotlikoff and Scott Burns, *The Coming Generational Storm: What You Need to Know About America's Economic Future.* Cambridge, Massachusetts: MIT Press, 2005.

Paul Krugman, *The Return of Depression Economics and The Crisis of 2008*. New York: W.W. Norton, 2009.

Joel Kurtzman, *The Death of Money: How the Electronic Economy Has Destabilized the World's Markets and Created Financial Chaos.* New York: Simon & Schuster, 1993.

Arthur B. Laffer, Stephen Moore and Peter J. Tanous, *The End of Prosperity: How Higher Taxes Will Doom the Economy – If We Let It Happen.* New York: Threshold Editions / Simon & Schuster, 2008.

Arthur B. Laffer and Stephen Moore, *Return to Prosperity: How America Can Regain Its Economic Superpower Status.* New York: Threshold Editions / Simon & Schuster, 2010.

John Lanchester, *I.O.U.: Why Everyone Owes Everyone and No One Can Pay.* New York: Simon & Schuster, 2010.

David S. Landes, *The Wealth and Poverty of Nations: Why Some Are So Rich and Some So Poor.* New York: W.W. Norton, 1998.

Lewis E. Lehrman, *The True Gold Standard – A Monetary Reform Plan Without Official Reserve Currencies*. Greenwich, Connecticut: The Lehrman Institute, 2011.

Louise Levathes, *When China Ruled the Seas: The Treasure Fleet of the Dragon Throne, 1405-*

1433. Oxford: Oxford University Press, 1997.

Michael Lewis, *Panic: The Story of Modern Financial Insanity.* New York: W.W. Norton, 2009.

Naphtali Lewis and Meyer Reinhold (Editors), *Roman Civilization: Sourcebook II: The Empire.* New York: Harper Torchbooks, 1966.

Nathan Lewis and Addison Wiggin, *Gold: The Once and Future Money.* Hoboken, New Jersey: John Wiley & Sons, 2007.

Charles A. Lindbergh, Sr., *Lindbergh on the Federal Reserve* (Formerly titled: *The Economic Pinch*). Costa Mesa, California: Noontide Press, 1989.

Deirdre N. McCloskey, *Bourgeois Dignity: Why Economics Can't Explain the Modern World.* Chicago: University of Chicago Press, 2010.

Heather MacDonald, *The Burden of Bad Ideas: How Modern Intellectuals Misshape Our Society.* Chicago: Ivan R. Dee, 2000.

Bethany McLean and Joe Nocera, *All the Devils Are Here: The Hidden History of the Financial Crisis.* New York: Portfolio/Penguin, 2010.

Karl Marx and Friedrich Engels, *The Communist Manifesto*. London: Penguin Classics, 1985.

John Mauldin and Jonathan Tepper, *Endgame: The End of the Debt Supercycle and How It Changes Everything.* Hoboken, New Jersey: John Wiley & Sons, 2011.

Martin Mayer, *The Fed: The Inside Story of How the World's Most Powerful Financial Institution Drives the Markets.* New York: Free Press, 2001.

Michael Medved, *The 5 Big Lies About American Business: Combating Smears Against the Free-Market Economy.* New York: Crown Forum, 2009.

David I. Meiselman and Arthur B. Laffer (Editors), *The Phenomenon of Worldwide Inflation.* Washington, D.C.: American Enterprise Institute, 1975.

Gavin Menzies, *1421: The Year China Discovered America.* New York: Harper Perennial, 2002.

____________, *1434: The Year a Magnificent Chinese Fleet Sailed to Italy and Ignited the Renaissance.* New York: Harper Perennial, 2009.

Brendan Miniter (Ed.), *The 4% Solution: Unleashing the Economic Growth America Needs.* New York: Crown Business / George W. Bush Institute, 2012.

Hyman P. Minsky, *John Maynard Keynes*. New York: McGraw-Hill, 2008.

______________, *Stabilizing an Unstable Economy.* New York: McGraw-Hill, 2008.

Ludwig von Mises, *The Anti-Capitalist Mentality.* Princeton, New Jersey: D. Van Nostrand Company, 1956.

_______________, *Human Action: A Treatise on Economics.* Third Revised Edition. Chicago: Contemporary Books, 1966.

_______________, *On the Manipulation of Money and Credit.* Dobbs Ferry, New York: Free Market Books, 1978.

_______________, *The Theory of Money and Credit*, New Edition. Irvington-on-Hudson, NY: Foundation for Economic Education, 1971.

Stephen Moore, *How Barack Obama Is Bankrupting the U.S. Economy* (Encounter Broadside No. 4). New York: Encounter Books, 2009.

Charles R. Morris, *The Trillion Dollar Meltdown: Easy Money, High Rollers, and the Great Credit Crash.* New York: Public Affairs/Perseus, 2008.

Robert P. Murphy, *The Politically Incorrect Guide to Capitalism*. Washington, D.C.: Regnery, 2007.

Charles Murray, *Coming Apart: The State of White America, 1960-2010*. New York: Crown Forum, 2012.

____________, *What It Means to Be a Libertarian: A Personal Interpretation*. New York: Broadway Books, 1997.

Andrew P. Napolitano, *Lies the Government Told You: Myth, Power, and Deception in American History.* Nashville: Thomas Nelson, 2010.

Sylvia Nasar, *Grand Pursuit: The Story of Economic Genius*. New York: Simon & Schuster, 2011.

Maxwell Newton, *The Fed: Inside the Federal Reserve, the Secret Power Center that Controls the American Economy.* New York: Times Books, 1983.

Johan Norberg, *Financial Fiasco: How America's Infatuation with Home Ownership and Easy Money Created the Economic Crisis*. Washington, D.C.: Cato Institute, 2009.

Grover Norquist and John R. Lott, Jr., *Debacle: Obama's War on Jobs and Growth and what We Can do Now to Regain Our Future.* New York: Wiley, 2012.

Mancur Olson, *The Logic of Collective Action: Public Goods and the Theory of Groups*, Revised Edition. Cambridge, Massachusetts: Harvard University Press, 1971.

___________, *Power and Prosperity: Outgrowing Communist and Capitalist Dictatorships*. New York: Basic Books, 2000.

___________, *The Rise and Decline of Nations: Economic Growth, Stagflation, and Social Rigidities*. New Haven, Connecticut: Yale University Press, 1984.

Ron Paul, *End The Fed*. New York: Grand Central Publishing / Hachette, 2009.

_______, *Liberty Defined: 50 Essential Issues That Affect Our Freedom.* New York: Grand Central Publishing / Hachette, 2011.

_______, *Pillars of Prosperity: Free Markets, Honest Money, Private Property.* Ludwig von Mises Institute, 2008.

_______, *The Revolution: A Manifesto*. New York: Grand Central Publishing / Hachette, 2008.

Ron Paul and Lewis Lehrman, *The Case for Gold: A Minority Report of the U.S. Gold Commission.* Ludwig von Mises Institute, 2007. This can be downloaded from the Internet at no cost from http://mises.org/books/caseforgold.pdf

Peter G. Peterson, *Running On Empty: How the Democratic and Republican Parties Are Bankrupting Our Future and What Americans Can Do About It.* New York: Farrar, Straus and Giroux, 2004.

Kevin Phillips, *Bad Money: Reckless Finance, Failed Politics, and the Global Crisis of American Capitalism*. New York: Viking Press, 2008.

___________, *Boiling Point: Democrats, Republicans, and the Decline of Middle-Class Prosperity.* New York: Random House, 1993.

Lowell Ponte, *The Cooling*. Englewood Cliffs, New Jersey: Prentice-Hall, 1976.

Richard A. Posner, *The Crisis of Capitalist Democracy*. Cambridge, Massachusetts: Harvard University Press, 2010.

_______________, *A Failure of Capitalism: The Crisis of '08 and the Descent into Depression.* Cambridge, Massachusetts: Harvard University Press, 2009.

Virginia Postrel, *The Future and Its Enemies: The Growing Conflict Over Creativity, Enterprise, and Progress.* New York: Free Press, 1998.

Raghuram G. Rajan, *Fault Lines: How Hidden Fractures Still Threaten the World Economy.* Princeton, New Jersey: Princeton University Press, 2010.

Joshua Cooper Ramo, *The Age of the Unthinkable: Why the New World Disorder Constantly Surprises Us And What We Can Do About It*. New York: Little Brown / Hachette, 2009.

Ayn Rand, *Capitalism: The Unknown Ideal (With additional articles by Nathaniel Branden, Alan Greenspan, and Robert Hessen)*. New York: Signet / New American Library, 1967.

Carmen M. Reinhart and Kenneth S. Rogoff, *This Time Is Different: Eight Centuries of Financial Folly*. Princeton, New Jersey: Princeton University Press, 2009.

James Rickards, *Currency Wars: The Making of the Next Global Crisis*. New York: Portfolio/Penguin, 2011.

Barry Ritzholtz with Aaron Task, *Bailout Nation: How Greed and Easy Money Corrupted Wall Street and Shook the World Economy*. Hoboken, New Jersey: John Wiley & Sons, 2009.

Wilhelm Roepke, *A Humane Economy: The Social Framework of the Free Market*. Chicago: Henry Regnery Company, 1960. This can be downloaded from the Internet at no cost from http://mises.org/books/Humane_Economy_Ropke.pdf

Murray N. Rothbard, *America's Great Depression*. Fifth Edition. Auburn, Alabama: Ludwig von Mises Institute, 2000. This can be downloaded from the Internet at no cost from http://mises.org/rothbard/agd.pdf

________________, *The Case Against the Fed.* Second Edition. Auburn, Alabama: Ludwig von Mises Institute, 2007. A version of this book can be downloaded from the Internet at no cost from http://mises.org/books/Fed.pdf

________________, *A History of Money and Banking in the United States: The Colonial Era to World War II.* Auburn, Alabama: Ludwig von Mises Institute, 2002. This can be downloaded from the Internet at no cost from http://mises.org/Books/HistoryofMoney.pdf

________________, *The Mystery of Banking*. Second Edition. Auburn, Alabama: Ludwig von Mises Institute, 2008. This can be downloaded from the Internet at no cost from http://mises.org/Books/MysteryofBanking.pdf

________________, *What Has Government Done to Our Money?* Auburn, Alabama: Ludwig von

Mises Institute, 2008. This can be downloaded from the Internet at no cost from http://mises.org/Books/Whathasgovernmentdone.pdf

__________________, *For a New Liberty: The Libertarian Manifesto* (Revised Edition). New York: Collier Books / Macmillian, 1978.

Michael Rothschild, *Bionomics: The Inevitability of Capitalism.* New York: John Macrae / Henry Holt and Company, 1990.

Nouriel Roubini and Stephen Mihm, *Crisis Economics: A Crash Course in the Future of Finance.* New York: Penguin Books, 2010.

Robert J. Samuelson, *The Good Life and Its Discontents: The American Dream in the Age of Entitlement 1945-1995*. New York: Times Books, 1995.

__________________, *The Great Inflation and Its Aftermath: The Transformation of America's Economy, Politics and Society.* New York: Random House, 2008.

Peter D. Schiff and Andrew J. Schiff, *How an Economy Grows and Why It Crashes.* Hoboken, New Jersey: John Wiley & Sons, 2010.

Detlev S. Schlichter, *Paper Money Collapse: The Folly of Elastic Money and the Coming Monetary Breakdown.* New York: John Wiley & Sons, 2011.

Robert L. Schuettinger and Eamonn F. Butler, *Forty Centuries of Wage and Price Controls: How NOT to Fight Inflation.* Washington, D.C.: Heritage Foundation, 1979. This can be downloaded from the Internet at no cost from http://mises.org/books/fortycenturies.pdf

Barry Schwartz, *The Paradox of Choice: Why More Is Less.* New York: Ecco / Harper Collins, 2004.

George Selgin and others, *Has the Fed Been a Failure?* Revised Edition. (Monograph). Washington, D.C.: Cato Institute, 2010.

Hans F. Sennholz, "Inflation Is Theft," *LewRockwell.com*, June 24, 2005. URL: http://www.lewrockwell.com/orig6/sennholz6.html

Hans F. Sennholz (Editor), *Inflation Is Theft*. Irvington-on-Hudson, New York: Foundation for Economic Education, 1994. A copy of this book may be downloaded at no cost from FEE's website at http://fee.org/wp-content/uploads/2009/11/InflationisTheft.pdf

Amity Shlaes, *The Forgotten Man: A New History of the Great Depression.* New York: Harper Collins, 2007.

___________, *The Greedy Hand: How Taxes Drive Americans Crazy And What to Do About It.* New York: Random House, 1999.

Julian L. Simon, *The Ultimate Resource.* Princeton, New Jersey: Princeton University Press, 1981.

Mark Skousen, *Economics of a Pure Gold Standard.* Seattle: CreateSpace, 2010.

____________, *The Making of Modern Economics: The Lives and Ideas of the Great Thinkers.* Second Edition. Armonk, New York: M.E. Sharpe, 2009.

Craig R. Smith, *Rediscovering Gold in the 21st Century.* Sixth Edition. Phoenix: Idea Factory Press, 2007.

____________, *The Uses of Inflation: Monetary Policy and Governance in the 21st Century* (Monograph). Phoenix: Swiss America Trading Company, 2011.

Craig R. Smith and Lowell Ponte, *Crashing the Dollar: How to Survive a Global Currency Collapse.* Phoenix: Idea Factory Press, 2010.

___________________________, *The Inflation Deception: Six Ways Government Tricks Us...And Seven Ways to Stop It!* Phoenix: Idea Factory Press, 2011.

___________________________, *Re-Making Money: Ways to Restore America's Optimistic Golden Age.* Phoenix: Idea Factory Press, 2011

Guy Sorman, *Economics Does Not Lie: A Defense of the Free Market in a Time of Crisis.* New York: Encounter Books, 2009.

George Soros, *The Age of Fallibility: Consequences of the War on Terror.* New York: Public Affairs, 2007.

___________, *The Bubble of American Supremacy: the Cost's of Bush's War in Iraq.* London: Weidenfeld & Nicolson, 2004.

___________, *George Soros on Globalization.* New York: Public Affairs, 2005.

___________, *The New Paradigm for Financial Markets: The Credit Crisis of 2008 and What It Means.* New York: Public Affairs, 2008.

___________, *Open Society: Reforming Global Capitalism.* New York: Public Affairs, 2000.

___________, *The Soros Lectures at the Central European University.* New York: Public Affairs, 2010.

Thomas Sowell, *Basic Economics: A Common Sense Guide to the Economy.* Third Edition. New York: Basic Books / Perseus, 2007.

_____________, *Dismantling America.* New York: Basic Books, 2010.

_____________, *Economic Facts and Fallacies.* Second Edition. New York: Basic Books, 2011.

_____________, *The Housing Boom and Bust.* Revised Edition. New York: Basic Books, 2010.

_____________, *On Classical Economics.* New Haven, Connecticut: Yale University Press, 2007.

Mark Steyn, *After America: Get Ready for Armageddon.* Washington, D.C.: Regnery, 2011.

_________, *America Alone: The End of the World As We Know It.* Washington, D.C.: Regnery, 2008. *[Full Disclosure: Steyn quotes Lowell Ponte in this book.]*

Joseph E. Stiglitz, *Freefall: America, Free Markets, and the Sinking of the World Economy.* New York: W.W. Norton, 2010.

_______________, *Globalization and Its Discontents.* New York: W.W. Norton, 2002.

Paola Subacchi and John Driffill (Editors), *Beyond the Dollar: Rethinking the International Monetary System.* London: Chatham House / Royal Institute of International Affairs, 2010. URL: http://www.chathamhouse.org/sites/default/files/public/Research/International%20Economics/r0310_ims.pdf

Ron Suskind, *Confidence Men: Wall Street, Washington, and the Education of a President.* New

York: Harper Collins, 2011.

Nassim Nicholas Taleb, *The Black Swan: The Impact of the Highly Improbable.* Second Edition. New York: Random House, 2010.

Peter J. Tanous and Jeff Cox, *Debt, Deficits and the Demise of the American Economy.* Hoboken, New Jersey: John Wiley & Sons, 2011.

Johan Van Overtveldt, *Bernanke's Test: Ben Bernanke, Alan Greenspan and the Drama of the Central Banker.* Chicago: B2 Books/Agate Publishing, 2009.

Damon Vickers, *The Day After the Dollar Crashes: A Survival Guide for the Rise of the New World Order.* Hoboken, New Jersey: John Wiley & Sons, 2011.

William Voegeli, *Never Enough: America's Limitless Welfare State*. New York: Encounter Books, 2010.

M.W. Walbert, *The Coming Battle: A Complete History of the National Banking Money Power in the United States*. Chicago: W.B. Conkey Company, 1899. Reprinted by Walter Publishing & Research, Merlin, Oregon, 1997.

David M. Walker, *Comeback America: Turning the Country Around and Restoring Fiscal Responsibility*. New York: Random House, 2009.

Jude Wanniski, *The Way the World Works*. New York: Touchstone / Simon & Schuster, 1978.

Jack Weatherford, *The History of Money: From Sandstone to Cyberspace*. New York: Crown Publishers, 1997.

Carolyn Webber and Aaron Wildavsky, *A History of Taxation and Expenditure in the Western World.* New York: Simon & Schuster, 1986.

Janine R. Wedel, *Shadow Elite: How the World's New Power Brokers Undermine Democracy, Government and the Free Market.* New York: Basic Books / Perseus, 2009.

Eric J. Weiner, *The Shadow Market: How a Group of Wealthy Nations and Powerful Investors Secretly Dominate the World.* New York: Scribner, 2010.

R. Christopher Whalen, *Inflated: How Money and Debt Built the American Dream.* Hoboken, New Jersey: John Wiley & Sons, 2010.

Lawrence H. White, *Is The Gold Standard Still the Gold Standard among Monetary Systems?* (Monograph). Washington, D.C.: Cato Institute, February 8, 2008. URL: http://www.cato.org/pubs/bp/bp100.pdf

Peter C. Whybrow, *American Mania: When More Is Not Enough*. New York: W.W. Norton, 2005.

Addison Wiggin and William Bonner, *Financial Reckoning Day Fallout: Surviving Today's Global Depression.* Hoboken, New Jersey: John Wiley & Sons, 2009.

Addison Wiggin and Kate Incontrera, *I.O.U.S.A.: One Nation. Under Stress. In Debt.* Hoboken, New Jersey: John Wiley & Sons, 2008.

Aaron Wildavsky, *How to Limit Government Spending...*, Berkeley, California: University of California Press, 1980.

John Williams, *Hyperinflation 2012: Special Commentary Number 414*. Shadow Government Statistics (Shadowstats), January 25, 2012. URL: http://www.shadowstats.com/article/no-414-hyperinflation-special-report-2012

Jonathan Williams (Editor), *Money: A History*. New York: St. Martin's Press, 1997.

David Wolman, *The End of Money: Counterfeiters, Preachers, Techies, Dreamers – and the Coming Cashless Society.* Boston: Da Capo Press / Perseus Books, 2012.

Thomas E. Woods, Jr., *Meltdown: A Free-Market Look at Why the Stock Market Collapsed, the Economy Tanked, and Government Bailouts Will Make Things Worse.* Washington, D.C.: Regnery Publishing, 2009.

__________________, *Nullification: How to Resist Federal Tyranny in the 21st Century.* Washinton, D.C.: Regnery Publishing, 2010.

__________________, *Rollback: Repealing Big Government Before the Coming Fiscal Collapse.* Washington, D.C.: Regnery Publishing, 2011.

Thomas E. Woods, Jr., and Kevin R.C. Gutzman, *Who Killed the Constitution?: The Federal Government vs. American Liberty From WWI to Barack Obama.* New York: Three Rivers Press, 2009.

Bob Woodward, *Maestro: Greenspan's Fed and the American Boom.* New York: Simon & Schuster, 2000.

______________, *The Power of Politics.* New York: Simon & Schuster, 2012.

L. Randall Wray, *Modern Money Theory: A Primer on Macroeconomics for Sovereign Monetary Systems.* London: Palgrave Macmillan, 2012.

______________, *Understanding Modern Money: The Key to Full Employment and Price Stability.* Northampton, Massachusetts: Edward Elgar Publishing, 2006.

Fareed Zakaria, *The Future of Freedom: Illiberal Democracy at Home and Abroad.* New York: W.W. Norton, 2003.

______________, *The Post-American World.* New York: W.W. Norton, 2009.

Luigi Zingales, *A Capitalism for the People: Recapturing the Lost Genius of American Prosperity.* New York: Basic Books/Perseus Books, 2012.